1974 BENJAMIN F. FAIRLESS MEMORIAL LECTURES

Library of Congress Catalog Card Number 75-12001
ISBN Number 0-915604-03-5

The Washington Embrace of Business

Roger M. Blough

Distributed by Columbia University Press

New York-London

The Benjamin F. Fairless Memorial Lectures endowment fund has been established at Carnegie-Mellon University to support an annual series of lectures. An internationally known figure from the worlds of business, government, or education is invited each year to present three lectures at Carnegie under the auspices of its Graduate School of Industrial Administration. In general, the lectures will be concerned with some aspects of business or public administration; the relationships between business and government, management and labor; or a subject related to the themes of preserving economic freedom, human liberty, and the strengthening of individual enterprise—all of which were matters of deep concern to Mr. Fairless throughout his career.

Mr. Fairless was president of United States Steel Corporation for fifteen years, and chairman of the board from 1952 until his retirement in 1955. A friend of Carnegie-Mellon University for many years, he served on the board of trustees from 1952 until his death. In 1959 he was named honorary chairman of the board. He was also a leader and co-chairman of Carnegie-Mellon's first development program, from its beginning in 1957.

Roger M. Blough, now a partner in the law firm of White & Case, was chairman of the Board of Directors and chief executive officer of United States Steel Corporation from May 3, 1955 through January 31, 1969. He holds a B.A. from Susquehanna University, and a Juris Doctor from Yale Law School. While at Yale, he was an editor of the YALE LAW JOURNAL. After graduation he engaged in the general practice of law with the firm of White & Case in New York City.

Mr. Blough was one of a group of attorneys engaged by United States Steel during the investigation of the steel industry by the Temporary National Economic Committee in 1939 and 1940. He was employed by United States Steel in February 1942 when he was appointed general solicitor in Pittsburgh in charge of all legal matters for the United States Steel Corporation of Delaware. In 1951 he became executive vice president-law, secretary, and director of the United States Steel Company which was formed through the merger of the Delaware corporation and three operating subsidiaries of United States Steel Corporation. United States Steel Company was merged into the corporation at the beginning of 1953 when the parent company became primarily an operating company.

Mr. Blough was elected vice chairman of the Board of Directors in May, 1952, and in August, 1952 was elected a director of United States Steel Corporation and a member of its Finance Committee.

Mr. Blough was vice chairman and general counsel of the corporation from 1953 until 1955, when he succeeded Benjamin F. Fairless as chairman of the Board of Directors and chief executive officer. In 1969, Mr. Blough retired from U. S. Steel and rejoined White & Case. He continues as a director of U. S. Steel, a member of the Finance Committee and Executive Committee, and a trustee of United States Steel Foundation, Inc.

He is a member of the Boards of Directors of The Equitable Life Assurance Society of the United States, Campbell Soup Company, Interpace Corporation, and Susquehanna University; a lifetime councillor, and former trustee and chairman of The Conference Board; a former chairman of The Business Council; an honorary director of The Commonwealth Fund; an honorary trustee of the Committee for Economic Development;

and a senior trustee of the United States Council of the International Chamber of Commerce, Inc.; and a trustee of The Presbyterian Hospital in the City of New York, and Grand Central Art Galleries.

He is the immediate past chairman of the Board of Directors of the Council for Financial Aid to Education, and a Fellow and the immediate past president of The Institute of Judicial Administration.

In 1972 Mr. Blough was appointed a member of the Commission on Executive, Legislative and Judicial Salaries by Chief Justice Warren E. Burger.

He assisted in the organization and was chairman of the Construction Users Anti-Inflation Roundtable until its merger in late 1972 with the Labor Law Study Committee to become The Business Roundtable of which he served as co-chairman.

A Word of Appreciation

The Benjamin F. Fairless Memorial Lectures were endowed in 1962 and begun the next year as a living tribute to an unusual American businessman.

It was a privilege which I greatly cherish to have been among those who worked with Ben Fairless from 1938 until his death in January 1962. Long an executive and a respected leader in the steel industry, Ben was, first and last, able, friendly, and companionable. Among his many interests were better corporate communications and he devoted many hours to this task within the corporation as well as in addressing audiences of all kinds.

He was also responsible for numerous public services including, importantly, some from which this great Carnegie-Mellon University benefited.

Born in modest circumstances, raised by an uncle and aunt, Ben Fairless had all the advantages of a humble beginning. He never lost sight of the commonplace and the virtues of finding practical, workable solutions.

He was clearly not a trained economist, but he spent a business lifetime in the midst of important economic matters. Never fully satisfied with his own contributions in this field of human endeavor, he undoubtedly would be gratified to have this annual series of lectures bear his name.

The relationship of government and business, the general subject of the current lectures, was a matter of continuing interest to Mr. Fairless.

1974

The Washington Embrace of Business

CHAPTER ONE / THIRTEEN
Washington Encirclement of Business Decisions

CHAPTER TWO / THIRTY-THREE
Economics and Politics in Conflict

CHAPTER THREE / FIFTY-ONE
"That Government Is Best . . . "

The Washington Embrace of Business

I. Washington Encirclement of Business Decisions

I. Washington Encirclement of Business Decisions

Trying to distill and communicate impressions of the last forty years about Washington and business relations with Washington is no mean task.

We must begin somewhere. So I start in 1933, in the midst of a severe depression, when Congress, in the early days of the first President Franklin D. Roosevelt administration, ended gold as the standard of U. S. currency, and adopted measures designed to reverse the depression, remedy unemployment, restore business confidence, as well as to curtail poverty and improve the social condition of the many in the nation.

The Beginning of Modern Day Controls

The National Recovery Act (NRA) was born, lived until 1935, and then died when the Supreme Court found the Act unconstitutional in the *Schechter Poultry* case. Following other adverse decisions including those involving the Bituminous Coal Conservation Act of 1935 and the Agriculture Adjustment Act of 1933, both of which involved serious and, as then determined, unconstitutional interventions by government, President Roosevelt made an all-out attempt to "pack," as it was called, the Supreme Court by increasing its number to fifteen. This 1937 effort was unsuccessful.

It seemed to some of us at the time that, even before the *Schechter* case, the NRA, with its attempts at controls of business and labor, was rapidly falling of its own intricate weight when the Supreme Court gave it the *coup de grace.*

But a number of the so-called "New Deal" measures of those times which were designed more effectively to control or benefit various elements of the economy did survive and are, in a sense, flourishing.

Legislation, however well intentioned, and whatever the morale lift it may give in trying times, rarely cures economic ills. So in spite of the legislative effort, throughout the thirties, bankruptcies and reorganizations multiplied along with serious unemployment. The brief business recovery of 1935 and 1936 was followed by a disastrous drop in the Fall of 1937 and early 1938.

It is not pleasant to recall that in 1938, just before the 1939 beginning in Europe of World War II, and after a vigorous thrust for six years of government interventions of one kind or another, the nation's unemployment reached one of its worst highs of about 19 per cent.

If legislated controls can motivate the economic success of the people as a whole, by any test of effectiveness that art must have eluded the legislators of the thirties.

Continuing our brief review of controls, which sometimes seem to have no practical alternative, in World War II they served many, although not always, useful purposes. World War II controls came to a decadent end, almost by general acclamation, in 1946. A similar but swifter demise terminated wage, price, and other controls after the Korean War in 1953. Memories are too fresh to require discussion of the peacetime wage and price controls begun in 1971 and ended in April of this year. And we now have, since August, a monitoring form of government surveillance.

Thus it might be concluded by those so minded that business, as well as labor and others, have successfully turned back the tide of government controls, short of war, and that therefore we can contend against and succeed in avoiding more government involvement in business decision making when we have the will to do so.

A part of my mission in these lectures is to point out why I think this view cannot be sustained.

Parenthetically, I seem to be addressing these remarks primarily to businessmen. Many other people have an equally keen interest in the why and the how of what business does, and of this I am quite aware. Those in other occupations, like the professions, the students and faculties of learning institutions, dozens of unions and union organizations, the organized churches, the increasing numbers of public sector employees, consumer groups, and hundreds of others, are vitally concerned and frequently vocal in just how Washington and business work out their mutual problems. Couching this statement in terms of business is not intended to neglect the presence or weight of the others or for that matter their respective predicaments. It is simply a convenience and one way of conserving time and effort in dealing with the subject matter, realizing the others can, if they care to, also supply a point of view, however critical. Moreover, they may perchance even find themselves in the same regulatory boat.

To continue, personally I believe too many American businessmen are still prone to the illusion that if the nation will just elect and appoint able and honorable men to high office in the legislative and administrative branches of government, Washington will, in major part, somehow go away and let business do unhampered what it knows best, the making and selling of products and the supplying of services. It calls to mind that famous line from one of the songs in the musical "South Pacific," "I'm gonna wash that man right out of my hair and send him on his way."

It is a good line for a musical but the businessman who thinks he can wash Washington "right out of his hair" is destined, I believe, for a painful surprise and much frustration.

One of the quests in this trilogy of lectures is to postulate why this is necessarily so and to formulate a few suggestions as to what to do about it.

Factors in the Growth of Government Controls

A number of reasons can be advanced for the growth and proliferation of government controls. The increasing specializations within society is one. Another is the remarkable increase in technological knowledge which in turn induces specialization. The marked change in numbers of people engaged in agriculture and the general urban movement can be listed. International considerations are a growing factor along with the important rise of multinational corporations.

Other factors include the mounting aspirations of nearly everyone for a better life, better medical care, education, housing, and means of transportation. To this may be added the desire of finding and financing ways to protect and provide for those who cannot fend for themselves.

In accomplishing these changes and programs, and the creation by fiat of economic benefits and the redistribution of income, law upon law, and regulation upon regulation have been enacted.

Not to be overlooked, although perhaps not to be overemphasized either, is the growth in population which our nation has experienced. With greater densities of customs and communities, and with the formalization of rules previously only informally observed, a multitude of new rules arise. Just what the exact relationship is between the number of people in any nation and the complexity of its regulatory system is hard to determine. But it is safe to say that as population grows and as people form themselves into operating groups such as businesses knowing that is the best and largely the only way to accomplish many forms of work, the whole system calls for more and more rule making bodies. We become more organized. Organization is rule making.

A Daniel Boone, alone in the woods with a long rifle, largely made his own rules until he came to a settlement at which point the joint rule making process began. And a Henry Ford producing an automobile was soon necessarily required to work with groups of people specializing in what they could do best. The groups multiplied in number and complexity as the products of 1900 gave way to the highway systems, the energy forms, the distribution centers, the competition of other automobiles, the import and export barriers, the multiple state and federal and other tax collection systems, the suppliers of thousands of parts, other groups specializing in metals, man made fibers, paints and coatings, air conditioners, radio and television systems, recording devices, credit and billing techniques, and many other specialized occupations.

This phenomenon of capital accumulation, massive production and social institutions can exist only in the presence of multitudes of people who in turn create the reality of markets for products, specialized and adapted to satisfy human known or anticipated needs and desires. And with it all, rule making flourished.

Yet I do not wish to attribute an undue amount of importance to population growth as an explanation of regulation growth. One need only to realize that if we stayed at 213 million population until the year 2000, we undoubtedly would still have an abundance of regulatory growth in those twenty-five years. On the other hand, if we had only one-tenth of our population today we would in all probability have fewer of all kinds of regulations than we now have. So population is an important factor.

seventeen

More people plus the complexities of society and the rise of aspirations tend, therefore, to create more rule making.

Twenty Years of Control Making

We need not pause in this sweep of analysis to do other than mention recent trends in the laws and legal requirements which have come to the successors of Daniel Boone and Henry Ford and to the rest of our nation's estimated 213 million population. If you will bear with me, I will mention some of the better known laws which have been enacted during the last twenty years by our Congress and which have had noticeable consequences from a control point of view. Their cumulative effect never ceases to amaze, or, should I say, startle me. I subject you to this only because of its therapeutic value.

Commodity Exchange Act of 1955
Defense Production Act Amendments of 1955
Fair Labor Standards Amendments of 1955
Bank Holding Company Act of 1956
Communications Act Amendments of 1956
Narcotic Control Act of 1956
Social Security Amendments of 1956
Civil Rights Act of 1957
Automobile Information Disclosure Act of 1958
Federal Aviation Act of 1958
Federal Boating Act of 1958
Flood Control Act of 1958
National Aeronautics and Space Act of 1958
Small Business Investment Act of 1958
Transportation Act of 1958
Welfare and Pension Plans Disclosure Act of 1958
Life Insurance Company Income Tax Act of 1959
Civil Rights Act of 1960
Federal Hazardous Substances Labeling Act of 1960
Land Acquisition Policy Act of 1960
Narcotics Manufacturing Act of 1960
Mineral Leasing Act Revision of 1960
Arms Control and Disarmament Act of 1961
Fair Labor Standards Amendments of 1961
Federal Water Pollution Control Act Amendments of 1961
Oil Pollution Act of 1961

Air Pollution Control Act of 1962
Antitrust Civil Process Act of 1962
Drug Amendments of 1962
Food and Agriculture Act of 1962
Revenue Act of 1962
Self-Employed Individuals Tax Retirement Act of 1962
Trade Expansion Act of 1962
Work Hours Act of 1962
Clean Air Act of 1963
Equal Pay Act of 1963
Civil Rights Act of 1964
Food Stamp Act of 1964
Interest Equalization Tax Act of 1964
Revenue Act of 1964
Securities Acts Amendments of 1964
Urban Mass Transportation Act of 1964
Automotive Products Trade Act of 1965
Federal Cigarette Labeling and Advertising Act of 1965
Food and Agricultural Act of 1965
Public Works and Economic Development Act of 1965
Voting Rights Act of 1965
Water Quality Act of 1965
Child Protection Act of 1966
Clean Air Act Amendments of 1966
Clean Water Restoration Act of 1966
Fair Packaging and Labeling Act of 1966
Fair Labor Standards Amendments of 1966
Federal Coal Mine Safety Act Amendments of 1966
Financial Institutions Supervisory Act of 1966
Oil Pollution of the Sea Act of 1966
Age Discrimination in Employment Act of 1967
Air Quality Act of 1967
Agricultural Fair Practices Act of 1968
Consumer Credit Protection Act of 1968
Natural Gas Pipeline Safety Act of 1968
Radiation Control for Health and Safety Act of 1968
Cigarette Smoking Act of 1969
Child Protection and Toy Safety Act of 1969
Federal Coal Mine Health and Safety Act of 1969
National Environmental Policy Act of 1969

nineteen

Tax Reform Act of 1969
Bank Holding Company Act Amendments of 1970
Bank Records and Foreign Transactions Act of 1970
Clean Air Amendments of 1970
Controlled Substances Act of 1970
Economic Stabilization Act of 1970
Environmental Quality Improvement Act of 1970
Fair Credit Reporting Act of 1970
Flood Control Act of 1970
Hazardous Materials Transportation Control Act of 1970
Investment Company Amendments Act of 1970
Mining and Minerals Policy Act of 1970
Noise Pollution and Abatement Act of 1970
Occupational Safety and Health Act of 1970
Rail Passenger Service Act of 1970
Resource Recovery Act of 1970
Securities Investor Protection Act of 1970
Water and Environmental Quality Improvement Act of 1970
Boat Safety Act of 1971
Economic Stabilization Act Amendments of 1971
Emergency Loan Guarantee Act of 1971
Export Administration Finance Act of 1971
Consumer Product Safety Act of 1972
Equal Employment Opportunity Act of 1972
Federal Election Campaign Act of 1972
Federal Environmental Pesticide Control Act of 1972
Federal Ship Financing Act of 1972
Marine Mammal Protection Act of 1972
Noise Control Act of 1972
Ports and Waterways Safety Act of 1972
Agriculture and Consumer Protection Act of 1973
Economic Stabilization Act Amendments of 1973
Emergency Petroleum Allocation Act of 1973
Federal Railroad Safety Authorization Act of 1973
Highway Safety Act of 1973
Oil Pollution Act Amendments of 1973
Fair Labor Standards Amendments of 1974
Water Resources Development Act of 1974

I have mentioned one hundred four significant new laws further regulating business during the past twenty years. In your thinking, supplement these laws with the thousands of rules, regulations, procedural requirements, and interpretations which necessarily accompany them and their innumerable predecessors passed prior to this recent twenty-year period. There are, in addition, uncountable governmental agency rules and regulations.

Bear in mind also that while federal civilian employment has remained virtually stable for six years at 2.7 million persons, total state and local government employment has risen to over 11 million. With this many employees, think fleetingly of the myriad of state and local rules and regulations coming out of all the political subdivisions they represent.

Moreover, as we are all aware, this year's crop of laws was bountiful. And the prospects are excellent for more new laws from the new Congress. To mention just a few: pension legislation enacted this year has a text seventy-five pages long. And now they begin to write the regulations. National health insurance is coming. Changes in the workman's compensation laws are being considered. The Office of Safety and Health Insurance has new regulatory proposals. And there is the continuous hazard of more attempts to contain inflation through wage and price controls and other means.

The point here is a simple one. Inviting as it may be to kick against the traces, that gesture will be futile if we hope by this means to avoid a growing myriad of controls of one sort or another.

I refer again to the four periods of specific controls of wages and prices during the past forty years. In all instances it may be said with some accuracy that businessmen, along with many others, had a hand in getting those controls lifted. Does this portend that we can somehow dampen down the course I have described of growth of laws and regulations? I personally believe the trend is as stated and dampening it will take some doing. For example, as this is written there is a strong undercurrent for reinstitution of more formal wage and price controls.

Ownership and Controls in Foreign Countries

Without wishing to equate world-wide government ownership of productive facilities with United States growth of regulation, as further evidence of the trends there seems to be an interesting comparability at least in time.

Lately we have seen a pronounced movement among foreign countries having raw materials either to assert more control of the further processing of their national resources or to assert outright ownership even if any of these steps involves violation of contracts. This is occurring in the oil industry in the Middle East and Latin America. It is also occurring in bauxite, iron ore, copper, and other minerals in Latin America, Africa, and elsewhere. In substance, these governments are asserting greater ownership rights, regardless of reason, over property which before belonged to domestic and foreign developers. Again, if these assertions are not of outright or partial ownership, they are of more regulation and controls and result in a far different division of the fruits of operation. There seems to be a blurring of that which was thought to be public and that believed to be private.

Developments in steel outside the United States confirm the trend. Out of a total estimated production of crude steel in 1973 of about 696 million metric tons, the communist countries, including China and North Korea, produced 209 million or 30 per cent of the total, and this, of course, came from government owned facilities. The International Iron and Steel Institute estimates another 88 million tons were produced by steel companies, in twenty-eight nations which were directly owned by their governments. Thus government production totals 43 per cent.

In addition, there are a number of other steel makers whose governments assert considerable control over their affairs. For example, in Japan, France, and the Benelux countries one wonders just how private their steel production is.

It is interesting to observe that the public sector's share in 1968 of total steel production of about 38 per cent moved up to an estimated 43 per cent in 1973. Also, that a recent survey made by International Iron and Steel Institute of steel plant projects in Europe and the developing countries, excluding United States and Canada, indicated 110 million metric tons of new capacity projected of which more than 90 million can be identified as public sector.

Minerals, oil, and steel are examples. Other selected industries can be found which will point to the same growth of more government control. And this is entirely aside from the important growth of public utilities.

Regulations Beget Regulations

It is that growth of rule making and regulation which concerns us. It may be put succinctly this way.

We are destined to have a rise in rule making. These rules grow even faster than the number of people to be regulated, almost geometrically in number and complexity. A demonstrable corollary is this. Business regulations beget more regulations and more regulators. With apologies to Parkinson, you may call that Blough's law if you wish.

Since our population grows, although more slowly, and since our dependence on a constantly rising number of new products grows apace, and since there is no end to our social aspirations, our rule makers are very likely to keep up. What some think of as a bureaucratic and compulsive drive by government personnel for importance and self preservation is an added and a mounting factor.

My concern is that if business continues its present stance, they, the rule makers, will exceed our gravest expectations in numbers of rules, in impact, and, in many cases, in lack of cogency.

Controls in Non-Business Activity

With controls growing in the business world, would not similar factors induce growth of more regulation for non-business groups and for individuals? A brief check will confirm that this is true. For example, all are aware of the recent legislation related to foundations which results in many more controls than previously existed and the end is not in sight.

As another example, in a sense the graduated income tax on individuals is a form of control. Although this form of taxation involves revenue raising, it quite definitely regulates income distribution.

Individuals also feel the effects of more controls in their daily activities, for example, health delivery costs. Without inferring merit or demerit to this relatively new form of regulation, recently there has been enacted comprehensive federal health planning, and much more extensive legislation is on the drawing boards. There are, for example, state requirements with respect to establishing the need for additional hospital rooms. In some cases the monitoring of admissions, the length of the stay of patients in hospitals, and even the establishment of fees which physicians charge for their services are now part of the control mechanism. Hospital and doctor bills are being audited in a number of instances in community health service programs. Let it also be noted that, with rising costs, the "voluntary" relinquishment of individual freedom of action in the health services area is welcomed by many.

Pragmatism as a Working Concept

Many other examples will occur to you of the embracing encirclement of regulations. Let us then give up our pointless and aimless thinking, and many of us have it, that we in business can curtail regulation by government, that we can wash that Uncle Sam right out of our hair.

Rather let us face up to the situation and adopt the approach of those who willingly share responsibility with government, of businessmen caught in the tide of long range regulatory currents, who travel, at times reluctantly, in the same inevitable direction as government, who seek similar ends after debating each one of those ends, and who supply a teamwork element and an experience within which good government and good business can survive to the redounding benefit of the populace.

With that as a working pragmatic concept, I would expect to find better rule making and, on balance, fewer rules.

Guidelines of Washington Cooperation

If the pragmatic concept of shared responsibility with government on specific proposals, after testing each on an individual merit basis, is to be followed, specifically how can business cooperate in joint actions with government and where does that road lead?

Take, for example, proximity to the action and to opinion forming. While concurring with the view that what the people in the nation outside of Washington think and say is what matters most in Washington, nevertheless opinion forming is a two-way street. What goes on in Washington is conveyed to the various constituents and tends to influence their thinking. Many national organizations, such as labor and consumer organizations, as well as the original research types, recognize this and are headquartered in Washington.

There are also, in Washington, a changing flow of ideas from younger minds, bright, adventurous, and iconoclastic. This younger group is constantly testing ideas in adversary forums. They help to purge the past and, while making their own mistakes, they make real contributions and have a broad influence on actions taken and on public opinion.

In a way Washington is the concentration point of all the heat pouring through the sun glass of public opinion. It is there the burning occurs and publicly appears.

So if business would follow the pragmatic concept, its presence must be felt in Washington even more than it now is.

It is elementary to say that Washington is demanding, organizationally and individually. The individuals selected in your corporate family to live in Washington and to specialize on its problems must be thin-skinned enough to hear and heed the constant signaling going on there, and at the same time be thick-skinned enough to take the criticisms that are so freely heaped on business generally and on business representatives.

Merely to survive in Washington is demanding. The circuit riding of Washington gatherings, its rumor mills, its gargantuan pressures from all directions, and its arms length suspicious relationships have special demands all their own. Notwithstanding all that, you and your associates will do well to have your presence properly represented in your capital city.

Business as an Expression of the People

In my view anyone who represents a business or groups of businesses in Washington need not be self conscious or move apologetically. Quite the contrary. I look at it this way.

A very high percentage of people gainfully employed in this country other than those who work for some branch of federal or state government are associated with one business or another. Corporate employment is about 55 per cent of all employment, but relatively few, other than government, employables are engaged in anything other than what would qualify under any broad definition as a form of production, professional, or service business, or auxiliary service connected with business. Therefore, what happens in business as broadly defined—the way people work, the things they do for government and ask their government to do for them, their style of life within business—is as much an expression of people, and as much a reality as a mural or a landscape. In the aggregate, business well represents a large proportion of the mosaic of American life. The forms differ and the satisfactions may be less or may be more, but in the broadest sense people in conducting businesses are expressing themselves. Their group action is the concrete evidence of that expression.

Government, also an expression of the people who constitute it and who elect it, wishes, I believe, to be as aware as possible of the expression of the millions whose lives are interwoven in the business world. In our form of representative government I know of no reason why leadership of business groups may not speak up in the interest of their own constituencies. In fact, it could well be termed an inadequacy in management if they do not. The unions do it, and special interest groups do it.

The fact that we have congressmen in Washington who are expected to represent the voters in their areas does not mean they have exclusivity of representation except with respect to the voting function. If so, union leaders or leaders of special interest groups would be precluded from taking their matters to the Hill. So I say that businessmen at all levels who represent the interests they serve need have no apologies for bringing their concerns and voicing their interest in Washington.

Some in Congress may resent or protest this and claim they alone represent all the people, including those gainfully employed in business. Whether their roll call votes constitute exclusive representation is, I suppose, a matter of definition. But recognizing that they do have an important role in representation is all the more reason for business leadership in their corporate representative capacity to make their views known.

Perhaps some antagonism between some on the side of government and some on the business side is to be expected. In the real world support of differing ideas or programs produces friction. This is not irreconcilable antagonism but a method of resolving differences. For business to assume a non-questioning approach to avoid controversy would be a serious loss in the finding of the better of two plans. It would lose the great gain to be had by full debate. Stalemates may occur but the gain from debate produces a value far in excess of any loss on this or that special project.

There are even a few left on the Hill or among consumer groups who talk about "big business interests" and infer public disaster or corruption when businessmen speak their minds. This *in personam* attack may be good politics but it certainly is no substitute for considering a suggestion on its merits and i believe most occupants of elected office on the Hill and their staffs will seek the meritorious answer.

Senior Executives in Washington

One of the perennial complaints that I hear relates to the infrequency with which top echelon executives visit the nation's capitol and the rule makers and regulators who constitute the Hill and the agencies of government. Amid all the other duties of a busy executive, it is not easy to find time for Washington. Moreover, the occasional success one has on some project there is likely to be forgotten amid the frequent disappointments. The drain on patience is enormous. But, nevertheless, I think it is part of the upper managers' job. After all, if you have followed the pragmatic concept up to now you will realize that you are talking with and meeting with those who have some management prerogatives which affect your company. What these government quasi-managers will do will likely affect your business as much or more than anything you can do for it directly.

A Business Executive's Biggest Job in Washington

Like everything else, it is the individual within the corporate group who can and does make the difference. Whether he came up in operations, in sales, or the financial end, or research, or any other part of the business, it is likely to be new territory for him when first he takes up his duties in Washington. One of his severest tasks will be to reorient his own thinking. One way of saying this is that your biggest job in Washington is you.

If you recognize that government has and will continue to have a place, probably an enlarging one, and has or is likely to get the authority it thinks is needed, then it will be easier to be successful in Washington. You have become a joint problem solver with government. While reserving your right to demonstrate why there is no problem, or that no new law or regulation is needed, you are likely to find that your specific solution type of approach is greatly appreciated.

The Reputation of Business in Washington

Many of us in the business and professional world are, I believe, super-sensitive and feel somewhat battered by the constant probings and public criticisms that seem to come the way of most business, especially the larger units and their leadership. Collaterally it has the effect of inducing or at least lending credence to laws intended to correct business insufficiencies.

While one can quarrel with some outlandish and uncalled for specific instances which constantly occur, on balance it has become a business expectation to find the media overemphasizing our shortcomings in their reporting of our doings. The fact that our shortcomings make the headlines is, in a way, a tribute to the solid daily accomplishments and conduct of most corporate units. In effect, it is a breach of the public's expectancy which gets publicized.

Regardless of how little or how much merit the various attacks may have, the hair shirt we wear has its benefits. This may sound to you like trying to make the best of an unfair situation. Perhaps it is. But there is a certain beneficial discipline in being criticized fairly or unfairly and being forced to grin and bear it.

This belief that criticism from without as well as from within our respective organizations is good for the soul is certainly not a novel idea. Lincoln reflected on his criticism in this manner:

> If I were to try to read, much less answer, all the attacks made on me, this shop might as well be closed for any other business. I do the very best I know how—the very best I can; and I mean to keep doing so until the end. If the end brings me out all right, what is said against me won't amount to anything. If the end brings me out wrong, ten angels swearing I was right would make no difference.[1]

It is true he was referring to the presidency and to the constant barrage of unfairness heaped upon his work there. But to a lesser degree the executive in almost any business must bear a taste of the same sort of thing.

If it gets too irresponsible, there are the laws against slander and libel. And I think it is well to remember that the rights

---

[1] Bartlett, *Familiar Quotations* (Little, Brown 1951), p 458.

preserved in the First Amendment of freedom of speech benefit all of us and, as a practical matter, would not exist if those of us in business were subject only to properly censored and just criticisms.

Public Communications

One concept which I support in the public communications field whether it be in Washington or elsewhere is to do a proper job, explain it, stand behind it, and take what comes. And this applies to taking a position on proposed legislation or on anything else. My view is that the quality of performance and the fairness of position is what counts. It is all too easy to get red in the face and shout, "Oh, the injustice of it all." when perhaps, in fact, some percentage of the criticism from the Hill or a consumer group was really justified due to our own inadequate performance.

The breast beating and *mea culpa* approach, however, can be overdone and has serious limitations. It implies that ordinary mortals like mere businessmen are justified in seeking perfection. All of us make so many errors there are plenty of occasions for us to say we were wrong and much about which we can and should be humble. But to claim the entire business community is all wrong or inadequate seems to me to be a futile exercise and in itself an unfair one.

I will not dwell here on the currently publicized crop of corporate misdeeds including such things as illegal contributions to political campaigns. And there are many, many pieces of unfinished corporate business such as clean air and water, better safety at the work place, higher quality research, or, simply, a better small car, or a house paint which will brace itself better against the elements, or fairer advertising, or, for that matter, another ten thousand steels to add to the first ten thousand. Here we are trying to concentrate on Washington and business relations. And that in itself is an ample subject.

The Social Role of Corporations

And while we are on the subject of communications with Washington and elsewhere, I would like to register a mild objection to all the time that is used in describing and justifying the so-called social role of corporations. Enough has been written on that subject to satisfy most of us for the next fifty years. It is not insensitivity to social needs to suggest that a part of this dialogue is largely a misinterpretation of basics. People compose society. Corporations are groups of individual people. Everyone, as a member of society, is involved in playing a social role of some kind whether at work or at play. The activity of citizens at work within the groups we call corporations obviously aggregate the major part of all social roles played by everyone in society. The question, therefore, is not, "Will the corporation play a 'social' role?" It is constantly playing a social role as an operating unit in society. The work purposes for which the individuals get together are accomplishing social ends and are as much social activities as slum clearance or fostering the Boy Scouts.

A more apt approach is to examine corporate performance in all its activities. Do the people in a particular corporate group provide a means of livelihood, a safe place to work, reasonable income for those in the group called employees, and fair employment practices, freedom of expression and internal communications, and environmental acceptability? Does group activity produce a product well designed and suited to its purpose, and is it sold at a competitive price? Are its customers fairly treated and do they have a reliable source of supply? Is its operation performed well enough to earn its competitive way and return to its investors a sufficient payment to justify their investment?

These social performances are at least as important as to ask, for example, whether the persons associated in an enterprise which we might call the Erstwhile Tool Company of North and South America were doing their part to educate blacks, as I, a self appointed critic, conceive that part to be? Or, for instance, may the Erstwhile Tool Company in good social conscience establish a subsidiary group in apartheid areas of Africa?

It is true that one type of so-called social activity of corporations may appeal more than another, and any one of us may say that this or that social purpose is a worthy one and another is less worthy. This evaluation applies to everything we do. Within it there is plenty of room for measuring equitable treatment of all who compose society and of giving full weight to human values.

And while we are considering worthy social purposes of a

group of individuals composing a corporation, let us not fail to emphasize that living a disciplined existence in a productive sense is a primary social good of first importance, the value of which is constantly appraised by other individuals who deal with the disciplined group. Those who acquire its products and services measure in a very practical way the social value of the group's performance.

So I am not too much concerned by criticism of the so-called lack of social role of corporations. I would, moreover, welcome an audit of social performances provided the audit is sufficiently comprehensive and objective.

In Washington or elsewhere those who represent the group need have no doubt of the importance of their role or the legitimacy of their presence or of their basic contribution to society.

How Will Corporations be Permitted to Operate?

Later we can discuss further what I believe to be about the most important role corporate representatives can play on the Washington scene. Suffice now to say that it involves a method of operation, or, if you prefer, a method of management for the long pull. What is the way that you and your government as sharers of responsibility in the production end of our society can best manage this important aspect of our national affairs? That, I think, is the large question. It is certainly not a simple one. But it is a question to which all of us knowingly or unknowingly devote much of our time. And we will come back to that later.

My effort in this first section of our discussion is to present a point of view, a sort of pragmatic concept. You and your associates will be more successful and your social obligations as producers will be better fulfilled if you approach your government on a shared responsibility basis. Undoubtedly this will be difficult to comprehend at both ends of the line. Your reception is bound to be mixed. But what salesman was successful all the time?

I repeat, that because of our history, our individual backgrounds, our natural inclinations as businessmen to want complete freedom and no red tape, to move expeditiously and to get the work done, the hardest job we have will be to sell ourselves.

thirty-one

There is, in fact, a Washington embrace. It will not go away. The question for Washington and for business is the ultimate workability of that embrace.

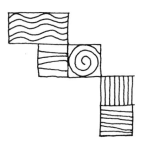

The Washington Embrace
of Business

II. Economics and Politics in Conflict

thirty-three

II. Economics and Politics in Conflict

In Washington the contest between economics and politics is never ceasing. And this is not an academic debate. It has many practical applications including a notable few in which I was involved.

Politics and Steel Prices

Steel prices have a curiously magnetic attraction for those occupying political positions. Let it be quickly said there is more historical than current reason for this. In earlier days the production and selling of steel had a much more pivotal position in the economic life of the nation than it does today.

As far back as Theodore Roosevelt, the office of the presidency was anxiously aware of the organization through merger of the United States Steel Corporation in 1901. For our purposes here, it is sufficient to say that initially U. S. Steel had about 66 per cent of the nation's steel production. This gradually fell until today it is about 23 per cent.

A government suit seeking U. S. Steel's dissolution on the ground of monopoly was begun in 1911 and successfully defended in the district and circuit federal courts. It lay dormant through World War I and, finally, in 1921, the Supreme Court, in a divided opinion, absolved U. S. Steel from that charge.

This was followed in 1924 by the Federal Trade Commission Basing Point case, a matter involving pricing and selling methods, which was resolved in 1951 with a consent order.

Without pausing to recount all of the vicissitudes of United States Steel and other steel makers over the years, let it only be noted that there were many in politics who took a special interest in steel making, in its labor relations, and especially in steel prices.

President Franklin Delano Roosevelt had explicit ideas regarding steel prices in 1938 which toward the end of a long depression period were reduced to a selling price approximating roughly where they were ten years earlier. The announcement said, "The price reductions are made to meet competitive conditions and with the hope that such reductions will stimulate a demand for steel products."

The Temporary National Economic Committee, authorized by Congress in 1938, held many hearings during 1939 and 1940.

Steel prices and various aspects of steel industry production and selling practices were thoroughly investigated. Begun during a period of very low production of steel, when it was supposed there was too much capacity, the hearings changed character with the advent of increased demand arising out of the commencement of World War II in Europe when it became apparent there was not enough United States steel capacity.

Steel Prices in World War II

The involvement of Washington in steel industry affairs rapidly became enlarged with the rise of World War II demands. In 1941 Leon Henderson, subsequently the head of the Office of Price Administration, called steel officials to Washington and, although without definitive authority, in substance denied them a price increase which undoubtedly would have followed a wage increase just granted and would have served in part to compensate for that increase. This squeeze on steel prices continued following Pearl Harbor and became intensified as wage and other costs rose during the war.

When President Harry Truman took over, one of his early 1946 tasks was to assist in resolving a labor dispute between the steel industry and the United Steelworkers of America. At that time World War II price controls were still in effect. Between 1941 and 1946 it had been expedient, I suppose one might say politically, to hold virtually unchanged the price of steel, but to grant through operation of the War Labor Board a number of wage cost increases in the steel industry, thus squeezing profits. Practically speaking, wage demands broke the camel's back of World War II rigid steel price controls. Under severe pressures from both sides and from prevailing circumstances, President Truman finally set the wage increase at 18½ cents per hour and allowed a price increase of $5 per ton.

The politics would have preferred a wage increase without a price increase; the economics required that both wages and prices be increased.

Senator Kefauver

Political interest in the economics of steel pricing resulted in incident after incident between 1946 and the middle fifties. Here I shall limit my comments to some steel industry contacts with Senator Estes Kefauver of Tennessee and also to an involvement with President John F. Kennedy. It is expected that a more detailed account of the Kennedy steel incident will be available later.

In the mid-fifties, Senator Kefauver, then chairman of the Judiciary Subcommittee on Antitrust and Monopoly of the Senate, was engrossed, it is fair to say, in matters economic which had a political tinge. Especially was he engrossed in steel prices. He was, with all this economic interest, concurrently engaged in advancing his candidacy for the office of the President, as, in fairness, were many others. We cannot fault Senator Kefauver for using the natural vantage point of senatorial hearings to emphasize political aspects which would enhance his chances of becoming his party's choice. This is, therefore, not to question Senator Kefauver's sincerity, when it is observed that the tone and content of his committee's proceedings were not exactly professionally dispassionate.

Much of the testimony given by officers of United States Steel, including myself, involved the economics of steel. It was said to be pivotal in the affairs of the nation. But putting it candidly, steel pricing was just too juicy a morsel from a political point of view to leave alone. I have no doubt the defenders of Senator Kefauver may construe those hearings quite differently. All I can say is that I am willing to let the record and the political setting speak for themselves.

It was a good illustration of how political motives, however justified, can relegate good economics to a secondary position. For vote appeal purposes, Senator Kefauver's attacks on large corporations were good politics. Many of the current attacks should likewise be examined from this point of view, although that is not to infer corporations should be above criticism. The inflationary effect of excessive government spending and of employment cost increases far above productivity improvement becomes obscured in the process.

Politically, this seems to serve a purpose. Actually, and ultimately, it is not good economics and not good for the economy as a profit starved industry will demonstrate. As I conceive it, over-stressing the role of steel prices or other prices, while it may be popular at the time, is therefore not good politics.

President John F. Kennedy

Being a constant football, at least in the past, steel labor negotiations and steel prices were also politically attractive to Senator John F. Kennedy who was later to become President. In the midst of the 116-day steel strike which began in the Fall of 1959, Senator Kennedy evidenced an interest in assisting the parties to the negotiations. At the time, President Eisenhower was out of the country and Vice President Richard M. Nixon was busily engaged in talking first with one side and then the other at his home in Washington in an effort to settle the strike.

Politically speaking, some credit was undoubtedly received by Vice President Nixon from the public for assisting in ending the grueling dispute. Politically speaking, it may also be said that this accomplishment did not bring comfort to any Democratic candidate for the presidency since it was widely believed Vice President Nixon would be the Republican candidate.

Wage Negotiations and Steel Prices

Steel prices were not involved in the 1959-60 labor negotiations. No price commitments were requested by Vice President Nixon and none was given in connection with the settlement of that strike when agreement was finally reached in January 1960. At this point a word about the market system and modern day labor negotiations is in order.

Competitive companies in the same industry may lawfully join together to negotiate a labor contract with a single union like the Steelworkers of America which represents in a number of steel producing companies substantially all wage earners engaged in steel making. But it would be illegal for the steel companies to agree about steel price increases even though the increases might be necessitated by higher labor costs. Therefore, from a fundamental and practical point of view, the antitrust laws can be strictly observed on the pricing side and must be, even though a number of competitive companies join together in negotiating a labor agreement. They join together in an attempt to more evenly balance the bargaining strength of multi-company unions.

It is vitally important, therefore, that whatever the government's desires may be regarding prices, the joint negotiations must be confined to wages and other non-price matters.

thirty-seven

This means that in the absence of government controls of prices and wages, prices are not a proper subject of discussion in any negotiation where the union represents employees of competing companies.

Against this background, as I said, and I wish to reemphasize this point, in the 116-day steel strike in 1959, no commitments were requested or given with respect to prices.

The background of possible changes in steel prices themselves was this. Nineteen hundred fifty-eight had been a recession year in steel. Nineteen hundred fifty-nine began as a low production year until the possibility of a strike built up steel demand. The pendency of the negotiation temporarily increased the demand for steel but the market outlook was only fair. Imports of steel were rising. Employment costs abroad were but a fraction of our own. Under these circumstances, my associates and I in United States Steel believed that market conditions would not competitively support higher steel prices.

So after the union had called a strike, in the hope of removing any doubts as to our corporation's intentions with respect to price I said on July 28, 1959 that,

> Whatever the length of the strike, and whatever the eventual outcome of the negotiations — so long as they are voluntary — we in United States Steel do not intend to raise the general level of our steel prices in the foreseeable future.

Wage increases had averaged about 8 per cent for two decades prior to the 1959-60 strike and that settlement involved a smaller increase of about 3½ per cent which represented a major change in trend. While the cost pressure from the wage increase itself was lessened, there was still considerable pressure but nevertheless there was no general price increase following the January 1960 settlement.

The 1962 Steel Price Incident

In the presidential election of November 1960, John Kennedy defeated Richard Nixon. In January of 1962 I met with President Kennedy in the White House along with the head of the Steelworkers and the Secretary of Labor. This meeting was on the initiative of the President.

When President Kennedy called us together he said in substance that he felt that our negotiation was of domestic and international importance. He spoke of a balanced budget in the same light. He said a steel inventory build-up with a subsequent let-down was not good. He said the country did not want a strike, that he hoped for an early settlement and one that would not involve a price change.

There were statements by the Secretary of Labor and by the head of the Steelworkers bearing on the negotiations.

For my part, I said I wanted to reaffirm my previous position that wage negotiations and prices were separate matters and that I did not wish to talk about prices with the union; that I appreciated the public interest referred to by the President, but in view of the costs previously incurred over a long period of time without price changes I did not want anyone present to be under the misapprehension of what the effect of additional costs might have in the industry.

I also said that there had been no dividend increases for five years, that capital expenditures were on the down grade, that debt had increased, that many projects were waiting for lack of money which would make the industry more competitive with Europeans and thereby provide more employment, that there were not enough earnings left over to have a strong steel industry and that corporate managers had to think of all phases of the business not only labor problems.

The negotiations started by mutual consent of the parties without any public urging of the President.

The labor settlement which eventually ensued was followed by an increase on April 10, 1962 of about 3½ per cent in the price of steel, first by United States Steel and then by a number of other steel companies. U. S. Steel's price announcement was followed by one of the most cataclysmic reactions by government to which any business or industry was ever subjected.

It is all too easy to get involved in the many ramifications of the Kennedy-steel industry incident. For our purposes here, we will try to limit the discussion to the conflict between what was good economics and what was considered to be good politics.

It must have seemed to President Kennedy to be politically desirable for him to slow the rate of inflation by keeping wage increases as low as possible and to play a part in avoiding another steel strike two years after the long one beginning in 1959, although the dangers of a steel strike in 1962 were minimal. It was also politically desirable to hope that steel prices would continue to be as stable as they had been for the prior four years. Good economics, however, called for a price increase, especially under circumstances of substantial wage cost increases and the need to modernize plants and equipment. Moreover, the demand for steel was such that the much needed price increase was thought by U. S. Steel to be attainable in the market place.

Incidentally, the labor settlement in 1946 was followed by a price increase as I have described. The same was true in 1947 and 1948. It was also true in 1949, in 1950, in 1952, in 1953, in 1955, in 1956, and in 1957. This was the only way the steel company could survive in the face of rising labor costs.

President Kennedy neither asked for nor was given by me or anyone else in United States Steel Corporation a commitment with respect to steel prices in connection with the 1962 wage negotiation. I have earlier mentioned his hope for prices and my response. My own feeling was then as it is now that this was not a proper matter for discussion in labor negotiations. It is still not clear how my various communications during that period could be construed as acquiescing in a no-price increase position.

Politically speaking, however, to have a labor settlement assisted by Vice President Nixon in early 1960 which was followed by no price increase, and then to have a labor settlement in 1962 made after a conference with President Kennedy to be followed by price increases, could appear to some to be politically discriminatory and to be a political setback however much a reasonable price increase would constitute "good" economics.

Essentially, this was clearly a major hang-up.

Certainly the President's desire to contain inflationary influences was unquestionably an important factor as the 1961 letters on steel prices indicated. But it was what President Kennedy felt to be a public image setback that seemed to me at the time to disturb him the most.

At his news conference on April 11, 1962, the day after our price increase, a reporter raised the question in these words:

Mr. President, two years ago, after the settlement [the steel labor settlement on January 4, 1960], I believe steel prices were not raised. Do you think there was an element of political discrimination in the behavior of the industry this year?

to which Mr. Kennedy answered,

I would not—and if there was, it doesn't really—if it was—if that was the purpose, that is comparatively unimportant to the damage that—the country's the one that suffers. I— If they do it in order to spite me it really isn't so important.[2]

There was also at the time some uninformed comment about a desire on our part to challenge the President. After four years of rising costs and no price increases, business judgment and competitive conditions dictated the action taken, not any desire to hurt the presidency or to assume a political or partisan role.

As most everyone knows, after great pressures from the White House, and after an extremely harsh presidential press conference, after a derogatory statement about businessmen which President Kennedy attributed to his successful business-man father, after many other statements and radio and tele-vision reports, plus reactions in Congress, FBI calls on news-paper reporters at early hours in the morning under the urging of Robert Kennedy, then Attorney General of the United States, after White House calls to executives of steel companies which had not yet increased their prices and to some which had, and after threats of grand jury proceedings, of cancellation of government contracts, and after a number of private conver-sations with steel executives and after several steel companies, led by Inland Steel Corporation, decided either not to increase their prices or to roll back their prices, United States Steel was also forced to rescind its projected price increase.

Economics and Steel Prices

Because we here examine the constant contest between good economics and good politics, it should be noted that while many in Congress supported the President, and while a number of newspapers and magazines and broadcasting media also approved his action, there was a sharp and very vocal opposi-tion, especially in the newspaper and magazine world.

A review of the April 11 to May 6 period of editorials, colum-nists' articles, cartoons, and letters, indicated the ratio was better than two to one favorable to U. S. Steel although the Corporation was severely criticized by many for the timing of

---

[2.] *The New York Times*, April 12, 1962, p 20.

the price increase, and the White House report on the communications received there showed them to be much more favorable to the President. Of interest is a letter exchange carried in *Time* magazine of May 4, 1962 when a reader observed,

I'll bet you a bottle of good Scotch that the letters you actually received ran at least two to one in favor of the President . . . .

Time's answer:

The drinks are on Reader Byrd. The letters ran more than five to one against President Kennedy's action.

The aftermath of the price confrontation was that in bits and pieces over the next year or year-and-a-half many of the price changes originally attempted were made, publicly announced and were generally sustained by the market. In every sense these changes were justified, and long overdue, and President Kennedy and his associates subsequently acted as though they recognized this. Economically speaking, although they had been the recipients of a serious political blow, companies in the steel industry slowly recovered to the still very low financial rates of return which lamentably have characterized the steel industry for some years before and some years after the 1962 incident.

The economics are simple from a cost and competitive point of view. When for twenty years improvement in steel shipments per man hour had amounted to only 2 per cent, with steel a capital intensive industry, and when the rate of profits in the steel industry was one of the lowest, if not the lowest, of any major industry, higher labor and other costs not compensated for by productivity over a period of four years had to be recovered, within competitive limitations, through price increases if the companies were to avoid further cost squeezes and continue to maintain equipment and compete in steel markets.

Steel is not the only industry that has fallen victim to the conflict between good economics and good politics which takes place periodically in Washington. Other instances could be enumerated, some in very recent days. But rather than discuss the specifics of other industries it is instructive to refer to the even larger conflict which arises in connection with the economic posture of the entire nation.

For example, total governmental expenditures for nine out of the last ten fiscal years (1964 through 1974) exceeded receipts. In only one year was there a surplus. Spending $102.9 billion more than the federal government took in was done for what appeared to be good political reasons. Whether or not justified, this, many believe, has been a major cause of the inflation which has resulted in the economic distress of the nation. Regardless of the possible benefits of deficit spending under a moderate Keynesian stance, when a nation overspends almost continuously during not only one but two decades, as ours did, something has to give. Consequently, we have seen, among other things, the dollar devalued twice and the price of gold soar.

"Good" Politics and Economic Limitations

This problem of government overspending is practically universal. Man's desires continuously exceed his reach. Our imaginations for seeking new governmentally financed programs for whatever good reasons, including the welfare of our fellow man, are virtually limitless. That is, they are limitless except for what might be termed our economic halter.

Any new program involves human effort for which some form of compensation is expected. This compensation must be obtained from some people and expended to others. To build the Verrazano-Narrows Bridge it was necessary to command the economic effort of many workmen who engaged in producing large quantities of cement and steel in great varieties, and hundreds of other materials, all requiring processing, transportation, communication and a myriad of other systems, as well as the payments made to thousands of owners for the temporary borrowing and utilization of their capital.

Paying these workmen at the site and compensating other workmen who supplied goods and services required large sums of money. Those who had saved temporarily loaned their savings in the expectation that future users of the bridge would repay the original capital borrowed.

This case of the Varrazano-Narrows Bridge is easily comprehended. It is used as an illustration because it has one major similarity with all other new programs whether they involve physical construction or otherwise. All programs require the time and talents of people and the resources of other people who make payment therefor. The compensation for these people, or, for instance, the payments made in welfare programs, must come from other people, mostly taxpayers or savers.

Since there are limitations both of people to do the work and of resources to pay them, there are necessarily waiting lines for the accomplishment of many proposals some of which must be deferred, however meritorious. This is true, whether the proposals be public or private. Whether the proposal involves a new bridge, or highway, or a project for new homes, or a school lunch program, or a social security system, or make-work activity in a depression, or a subsidy for the infirm, the answer is still the same. Each program has the effect of commanding the efforts and resources of people.

The problem of limitation is not confined to legislated programs. Any corporate executive knows of a number of highly meritorious projects which must compete with others of comparable merit. Every executive also knows that all the desirable things cannot be started at the same specific instant. Hence the scramble for position. Hence, also, the frequent over-borrowing to hurry the process and the problems which over-borrowing create.

In our discussions of the disciplinary effect of the price mechanism on decision making and on the allocation of resources, we consciously and naturally confine our remarks to the private sector. We have a tendency to assume that there is no similar disciplinary control in the public sector. But, as indicated, the public sector also has its own limitations although it does at times take longer for major breaches of those limitations to be exposed.

There are, of course, many differences between public and private proposals and public and private borrowing. But one thing they all have in common. They are necessarily bound by limitations in number and magnitude. This is legislative lesson number one. Yet it is one of the hardest to learn because most of us start out with the assumption that government is all-sufficient and all-powerful, which it is not. Government also has its limitations.

While we say deficit spending is a major source of inflation, it is only fair to examine the causes of the deficit spending and certainly the desire to aid our fellow man legislatively, even if it involves spending beyond our means, is one of those causes.

What we too seldom recognize is the extent to which government has tended to preempt resources leaving less for other sectors of the economy. This tends to limit that part of the resources available for reinvestment in productive facilities. This in turn gives rise to shortages and fewer goods than are needed which gives impetus to the demand-pull type of inflation.

The Rare Legislator

But even if limits to government spending and accomplishment are recognized, the game is only beginning. Whose project gets priority? That is where the real bargaining begins and rarely ends.

In the game of project priority, good economics frequently stands aside and what appears to be good politics sways the decision. After examining the feasibility of the project, its benefits, its cost in terms of people and compensation, and all the other factors which add up to good economics, our legislators frequently, and, I think to a degree necessarily, add in the ingredient of politics. Do my constituents want it? Will it be good for them? Will someone else give it to them if I don't? And will that someone else get the credit for giving it to them? What if it does create a deficit? We have had deficits before and lived through them. This special situation should have priority. It has, after all, the great merit of being "my" project. With election coming, a "good" legislator must do things for his constituents—things they can see or feel.

It is the rare legislator who can find a proper blend of economics and politics. It is so hard to sell the idea back home that it is good economics to allocate resources to another priority. This is probably too much to expect even from the elected superior and those concerned primarily with the public interest who compose our legislative bodies. But the real art remains. How to so allocate resources that both good government and good economics have a chance to survive.

And speaking of the arts, one of the arts we should curb or perhaps learn to do without altogether is the art of making promises to be paid for by others.

July 1974 figures show that transfer payments for government insurance, veterans' payments, old age, social security, welfare, and similar payments are now at the annual rate of $142.5 billion a year. Meritorious as the reasons are for such payments, their aggregate size is arresting and growing.

It is said that social security is heading for deep financial trouble. It is also said that we add to our obligations much faster than we discharge them. There must be a better way; but how to provide the legislative art to accomplish it?

To some extent this is happening. The recent legislative efforts of Congress, long advocated by business, to establish budgetary discipline is now law in the form of the Congressional Budget and Impoundment Control Act. While belated, it is, nevertheless, a highly salutary recognition of the role of good economics in good government. Its application has, however, yet to be tested.

Inflation and Politics

There are several more illustrations we could put on the table to show how this conflict between good economics and good politics operates.

It seemed to President Johnson to be good politics to push forward with the Great Society program with its overwhelming costs, simultaneously with incurring the enlarged and uncertain costs of the Vietnam War. It seemed to him also to be good politics not to ask Congress at that time to raise taxes to pay for these additional costs, a course which many of his closest advisers protested as not good economics. Inflation inevitably followed, although there is much more to the story than these few sentences.

Some of us were quite concerned about impending inflation of a serious order in 1969 and 1970. At that time it was not considered good politics to try to take certain wage restraint measures, especially in construction, when employment costs were a major source of the cost-push inflation then occurring. As it developed, however, it was a mistake not to regard inflation in its earlier stages as a serious political as well as an economic problem. With the advent of distorting controls and basic food cost increases, it soon became the number one political problem and still is. It became the number one problem politically in my view because it became the number one economic problem. In other words, there is no way in national policies to avoid an admixture of what we commonly call politics and economics.

Even if we define good politics as immediate response to a short range rumble or grumble, we cannot safely lose sight of the economics. Correcting the economics is the practical answer to reaching good political decisions.

A good example to illustrate the point is the recent action taken first by the President and then by Congress with respect to the report of the nine-man Commission on Executive, Legislative and Judicial Salaries. That report, weighed carefully by the appointed special Commission, recommended increases in Executive, Legislative and Judicial salaries which would have become effective in March 1974. This recommendation took into account a dry period of about five years (1969 to 1974) during which comparable salaries in lower government levels and in state administrations and in business were rising in the neighborhood of 30 per cent.

Following the Salary Commission's report, President Nixon, as the procedure provided, made recommendations to Congress. He reduced the Salary Commission's recommendations materially. Then members on the Hill, remembering that 1974 was an election year, pigeonholed even the reduced program. It was simply not good economics to do this, nor, I suspect, was it good politics.

The salaries of the higher grades in the Executive branch, frozen for five years, serve to suppress the grades immediately below them. This causes serious overlapping of lower salaries bumping against upper salaries. Legislators and judges are caught in the same economic tangle. Judges are being discriminated against in an inflationary period to the harm of the present judiciary and the prospects for obtaining good new appointees. Yet some legislators thought it was good politics to deny the increase.

To make the inequity more inequitable, in September more increases were authorized in the lower grades but none in the upper grades or in the Legislative or Judicial branches.

My own view is that the sooner we try to add in more economics and take out more politics in our legislative decision

making the better. But I recognize this is a bumpy road.
I also recognize that reasonable people may differ about what actually is good economics. The recent summit meeting in Washington and the pre-summit meetings which preceded it evidenced this fact. Divisions among the economists, labor, business, consumer spokesmen, and among the senators and representatives and others were not only apparent but educational.

Incidentally, this anti-inflation summit effort served valid purposes not the least of which was the experimental origination of a new method of reaching a basis for taking national economic action without necessarily achieving a consensus.

Is Politics on Trial?

Now, where does all this lead us? Well, for one thing, it should not lead to any attempt to submerge or outlaw so-called political pressure groups. Incidentally, I think they could better be called economic pressure groups. Let the farmers, or the unions, or the employers, or the government workers, or the exporters, or the importers, state their positions and vie for acceptance of those positions. Let the congressional committee which handles this or that specialty dwell upon, nurture and favor it, as they persistently do. Recognize the whole process for what it is—a highly motivated attempt to represent a position for constituents.

But at the same time, recognize that what seems to be the moment's good politics may actually have economic side effects which far outweigh any momentary political advantage. Reasonable and well informed capable people can and frequently will differ on what is good economics as I have said. This is to be expected, and calls for more debate, an accommodation of views, and, hopefully, a result that, economically speaking, can be lived with.

And it is only human to be motivated by political considerations. Positions, personal and party, are at stake. In major matters, however, good economics has a way on many occasions of constituting good politics. In fact, frequently the political solution is on trial where economics are concerned. And difficult decisions must be made. Witness the shipment of grain abroad in the face of shortages and higher grain and bread prices at home.

President Kennedy and the Pragmatic Principle

One more word about the Kennedy steel incident. While strong in political understanding, and quite knowledgeable about balance of payments determinants, President Kennedy was, to put it as fairly as possible, less experienced in business economics and such things as plain costs, prices, earnings and the effect of the absence of earnings. His personal reaction to the steel price increase was essentially political without comprehending that it was also a critical economic response which the massive declines in the stock markets and the reaction in the business world soon verified.

In the light of the business reaction which my private discussions with him after the original flare-up had forecast, President Kennedy was big enough to make overtures to the business community. In May of 1962, The Business Council readily responded. A committee, of which I was chairman, was appointed. It worked with him and had meetings with him during that summer, advising certain measures, especially in the area of taxation, consonant with actions he subsequently took and which, we believed, were beneficial to the economy.

On the other side of the coin, looking back twelve years later, it is undoubtedly true that those in steel, including myself, became so concerned by rising costs and so pressed to raise prices following the wage settlement, that we were not fully comprehensive of the serious political reaction President Kennedy and his associates were likely to have. Nor could we anticipate that any president of the United States would take or permit actions to be taken which would rouse such a wave of governmental and media hysteria that even President Kennedy became deeply concerned it would get out of hand.

All in all, it may be said that a greater awareness on both sides was warranted and would have served the public interest more adequately. Or, as the President said privately at the Economic Club dinner in New York on Friday, December 14, 1962, that both sides made a mistake; and that he was sorry it happened and that it must not happen again.

In a sense, less trauma might have resulted if we had given greater recognition to the cooperative pragmatic principle discussed in the first lecture.

Guardians of the Public Interest

Following the steel price incident, I wrote, in an article published in the January 29, 1963 issue of *Look* magazine,

> In closing, I would like to reflect briefly on 'the public interest,' a phrase that became part and parcel of the April struggle.
>
> What is the public interest? And who, if anyone, is the rightful custodian of the public interest?
>
> I believe that no one, however wise, however lofty in purpose, can exclusively define the public interest.
>
> I believe that each of us—teacher, minister, Government official, farmer, parent, editor, laborer, business owner, scientist, artist—sees the public interest from a different point of view.
>
> I thoroughly believe that Government officials, and legislators, have a right to define the public interest—as they see it. I think it is also proper for those charged with the management of a company to act with moral courage in what they genuinely consider to be in the best interests of their employers (the owners of the business), the employees and the customers, all of whom are part of the public. This is true because the operation of that business affects the health and growth of the American economy.
>
> But basically, 'the public interest' is the incredibly diverse lawful interest of each of more than 185 million Americans.
>
> I believe, therefore, that, among its many other responsibilities, it is the duty of Government to protect each of those individuals in the pursuit of his lawful interests, to insure that force and coercion do not intrude upon those interests, and to guarantee that each individual not intrude upon the lawful interests of others.
>
> Viewed in this light, I do not think it in the public interest in peacetime for anyone, including those of Presidential rank, to substitute his own action for the action of the marketplace by trying to set prices for any competitive products.
>
> Economic freedom, interlaced as it is with our spiritual and political freedoms, is precious and vital to the well-being of the American people. It must never be surrendered.

After more than a decade this still serves to convey my views as to who in America are guardians of the "public interest." This viewpoint bears on the necessity of melding good politics with good economics. It also reenforces the need for business to become a more integral part of the Washington scene and to offer its contribution in the policy developments which emanate from there. Although business has no corner on what is or is not good economics, its point of view should be helpful in assisting those who determine what is and what is not good politics.

The Washington Embrace
of Business

III. "That Government Is Best . . ."

III. "That Government Is Best . . ."

"That government is best which governs least" is a time honored and preceptive saying.

It may be paradoxical to point out, as I have, the logistics ahead of more rather than less government, of more rules and regulations and fewer unrestricted determinations by individual corporate or other groups than before, and at the same time to extoll the least government as the best government. Though it may be paradoxical, it is not contradictory. Just as it is not contradictory to say that the quantity of government regulation is no measure of its quality.

Likewise it may be paradoxical to observe that the regulatory mold within which business must operate is to a large degree dependent on business itself. Let us examine these postulates.

The Making of the Business Regulatory Mold

Congress and the regulatory agencies will, of course, make the mold of rules and regulations affecting business. But from what source will the ideas come? Perhaps it may be asserted with some assurance that those ideas will come largely from what may be termed the nation's intellectual community. This, of course, includes many now in government and in the universities. But any proper definition of the intellectual community must also include many of those engaged in business pursuits. By any test there is a wealth of intellectual initiative among businessmen.

Granting that the intelligence is there, and that business has made a valuable contribution, yet why have business ideas in general received relatively so little acceptance in the formulation of regulations affecting business in comparison, say, with ideas originating in the universities?

My own view is that this lack is not due to innate capacity so much as to two other factors. The first is the traditional approach of business to government which in substance says, "Please stay out of my affairs." The second is the plain lack of application of initiative, brain power, and manpower to this type of task.

Business Provides Initiative in Everything
Except Government Regulation

Most of us among businessmen dislike even to think of regulation until it strikes by interfering with what we are doing. My plea, if urging a trial of the pragmatic concept discussed before may be called a plea, is to try a new course, to get with the upcoming regulatory current of things and do something about it positively and specifically.

Not that I would expect following this concept to work miracles. Only that the nation and that massive segment of the nation represented by business occupations and activities will be much benefited if the new approach to government gradually gains ground.

Specifically what can be done? This is not an easy one to answer, but here is a starter.

The How of Business Initiative

Among its many initiatives, we relate here only to the broad subject of business initiative in the regulatory field.

One thing is an absolute *sine qua non*. A mountain-high amount of work is required, much more than business has believed necessary or productive before. Merely to identify the matters requires real effort. To name a few, take

—Safety at the working place.

—The better treatment of consumers.

—What happens to credit and an individual's personality in this computer age.

—Many phases of a fairly disagreeable set of transportation unsolvables in congested population areas.

—About a dozen aspects of inflation. Business participation in the recent summit meetings was a good start.

—Government absorption of business and competition with business.

—The too frequent willingness of the Hill to adopt, for political reasons, certain uneconomic union points of view.

—The shortages in natural resources.

—What to do about energy.

—Ecology.

—Economic education.

—Nondiscrimination.

—Sources of capital investment.

fifty-three

—How to compete internationally against a world of centrally controlled governments.

—How to offer a quality of life in a corporate world.

—The dismal absence on the Hill of appreciation of the effect on productive effort of that which they do.

Initiative is required on every one of these and many, many more matters, but who provides it?

Let us use an example for the sake of exposition. Consider the matter of requiring major corporations at first, and subsequently most corporations, to incorporate and operate under federal charters. The average businessman's reaction to this, mine included, is to discount its usefulness, magnify the inherent regulatory snags ahead, and suffer a traumatic mental blockage when the subject comes up as it has and will. It is currently suggested by some in Washington for the oil industry.

But is that the end of the matter? I think not. The point I make under the new approach is to open-mindedly study the objectives, the means, the probable domestic effects, the multinational values, the effects on the individual fifty states, the size and functioning feasibility of a new agency set up to operate federally charted corporations, and the costs or benefits to decentralized competitive operations as a way of life for this nation.

This is only one example of a "freeze" approach, or, by contrast, an opportunity for the pragmatic concept to operate. That is, there is here a problem to be solved. Let business and government get about it jointly and dispassionately.

The same may be said about co-determination about which we hear so much in Germany and elsewhere in Europe. Is it inapplicable, because of union opposition, to the United States and therefore may be forgotten? Or is it a matter of substance that needs study and resolution?

Organization for Regulatory Initiatives

Relations with government are as much a part of daily business as personnel policies or testing markets. It requires the attention of many people, some of whom must devote their full time to the effort. It also requires, in many instances, cooperative efforts with other business groups. Sometimes these take the form of industry associations or multiindustry associations such as, for example, the United States Chamber of Commerce or the National Association of Manufacturers.

It also requires specialist associations in research; for example, The Conference Board, or the Committee for Economic Development, or the Tax Foundation.

But it probably requires more than these. The need is for some form of association one of whose objectives is not only to originate but to advocate ideas wherever it can get a hearing, including the congressional Hill.

Washington Specialists

I believe the Washington job also requires talented specialists who are on the scene long enough to earn a respected place. It is true these specialists will be part of a corporation's overhead but a productive part as I see it, and how much more are corporate expenses increased by almost every regulatory offering out of Washington which affects business activity?

In substance there is now and in the future will be a real need for what may be called a "Washington professional." I use "professional" in the sense of one who practices an honorable occupation, whose conduct is above reproach, whose standards for conducting his daily work are known and open to inspection, who makes a career of professional representation, and who is frequently so much of an authority in a specialized field of regulation, for example pensions or quality of air, that his views earn a "public interest" position rather than a partisan reaction.

Earlier I suggested the usefulness of top echelon executives becoming real experts and spending major segments of their time in Washington. In a sense they are also part of the Washington professional team.

Washington–A Tough Testing Ground

It is hard work for those in Congress, in the regulatory agencies, and for those who compose the business corps and who are what I call Washington professionals.

Probably the hardest part is to uncover, debate, and present a "good" idea. It is equally difficult to examine the other fellow's "good" idea with the dispassionate study it deserves. One thing is certain. A poor, ill conceived idea is harder to present than a good one. And who wants to author a poor idea anyway?

fifty-five

So, much homework on Washington matters is necessary and, in the life of busy executives, hard to achieve. However much a company organizes to originate and present ideas, attention from those few composing the top echelon of management is vital. Any Washington idea must run the gauntlet on the Hill in numerous committees, and before or after that gauntlet it must run another with the Executive branch of government. But there is no good reason why business cannot come up with very workable ideas in the public interest. Some business organizations, such as CED, The Conference Board, and the Tax Foundation, have done this. On the whole, however, a disinterested count of initiatives actually achieved by business in Washington will turn up a relatively short list.

Yet this need not be so.

Just to recall, one of the premises with which we approach this matter is that regulations and government involvement in business decision making will not go away; that we are likely to have more intervention rather than less; that the valor of business had better be expended in the shared responsibility approach to governance of business, to obtaining rules with which it can live, to the avoidance of standards which smother business activity, than to expend such energy as it sometimes has in denouncing a regulatory agency or Congress, and then retreating into a secluded lair, all the better to lick its business wounds.

More United and More Positive

The purpose in this trilogy is to discuss trends in laws and regulations, attitudes, and, in particular, objectives. It is quite pertinent, therefore, to our inquiry, to consider the usefulness of one approach versus another. In this connection, the individual company approach—when regulations are being formulated—has usefulness but a rather limited one. The regulatory agency needs to consider the effects of its action on more than one company. If only one company suggests, or complains, it leaves the impression that other companies are not so adversely affected. It is not too much to say that perhaps more advanced and specialized joint mechanisms are needed by business in dealing with the advent of new regulatory agencies, or an issuance of a new set of regulations.

The Environmental Protection Agency, the Consumer Products Safety Commission, the Equal Employment Opportunity Commission, the Occupational Safety and Health Administration, to mention a few of the latter-day agencies, all have crucial impacts. To deal adequately with original concepts and initial regulations, something more than a simple company approach is usually needed. It probably requires some form of specialized, broadly representative, type of task force, knowledgeable of the work of the agency or program in question, and capable of presenting a rounded viewpoint representative of more than one industry or type of business activity. Again, we emphasize the attitude with which the task is approached. How can the best answer be found to our joint business-government problems, assuming a real problem exists?

Congressional Reaction to Business Ideas

Will such efforts by business be resented? Perhaps. But many in Congress will welcome and not resent disinterested, high quality assistance. Perhaps this effort would also aid in helping to allay some of the suspicion with which business is viewed by many in Congress.

I believe business can earn a respected place in the Washington scheme of things by doing at least three things for and with Congress:

—Initiate remedies for real problems that have a chance of being accepted and that will work. The word "initiate" is emphasized but a workable response to a real or even an imaginary ill is equally of value.

—Utilize an organization, probably one now existing, which, working with corporate executives who qualify as Washington professionals, can develop acceptable proposals on a variety of subjects.

—Methodically, unobtrusively but persistently and openly, convey these solutions to those in Congress for their consideration.

Some of this has been done in an *ad hoc* manner, but usually too little and too late. Much of what is done is of a defensive nature. A continued and searching solution-finding attitude by business is the answer. If this is done—carefully and thoughtfully—and presented by those having credibility, the organization accomplishing this rare feat would not only earn a right to initiate proposals but to comment knowledgeably on proposals made by others. Such comments, I believe, would likely be welcomed by many thoughtful elected and appointed men and women on the Hill.

The Congress

Congress is institutionally the most difficult branch of government with which to wrestle. It is, at once, the source of much which benefits the nation, and the source of much which troubles the nation.

Congress consists of a highly motivated group of elected representatives and senators, totaling five hundred thirty-five, most of whom possess unusual capabilities. It is characterized by a long, troubled, but, on the whole, remarkable history. It has a most intricate system of internal organization, committees, procedures and precedents. There are about seventeen standing committees of the Senate and twenty-one standing committees of the House. There are also a number of joint committees. In addition, there are over two hundred subcommittees. It is, and properly so, filled with strongly asserted individualism.

Although many senators and congressmen have had very long tenures, during the last five Congresses, including the newly elected at the beginning of the current one, there were four hundred one new names. One report indicates about one hundred new members were elected to Congress on November 5, this year. Each new individual means a new learning process. It means reviewing the past, examining hundreds of current matters, and, most importantly, learning the elements of how Congress can approach its business constituency, which, I repeat, is such a high proportion of all of us, to get the best results for the nation.

This process of institutional renewal in Congress is universal in all organizations, but the span of continuity of service varies greatly and this makes a world of difference. In passing, it may be noted that the turnover in the executive grades in the Executive branch of government is, from an organizational point of view, very high indeed.

Congress, like business, ranks close to the bottom in opinion poll surveys. Yet Congress is the chosen vehicle of the people in their search for satisfactory self-government. Unless Congress understands and develops an approach toward business, not to speak of other elements of society, which is conducive to the individual group system of production, operations and national productivity dependent on group operations are likely to deteriorate.

May Business "Look" at Congress?

It may be considered by some as presumptuous for business to look critically at Congress. Yet Congress is an essential part of our American tradition. Under this tradition the constituents of any organization may discuss its make-up and performance. People organized into large and small businesses and other groups are recipients of what is becoming more and more a matter of management by legislation and regulation, as I have indicated earlier. Why then, is it presumptuous to inquire and express viewpoints about the congressional part of the actual management of the entire system of American production and service?

To consider one example, it is common knowledge, as I have indicated, that when Congress spends more than it has for extended periods, inflation results. More than one businessman has railed against congressional spending as fostering inflation. As businessmen, we know what happens to a business that spends more than it has for too long. Yet business did relatively little on the Hill to help find a remedy for overspending in Congress.

It is true that the Committee for Economic Development has, in recent and past years, tackled some interesting problems relating to government spending. Also, in connection with congressional consideration of budget control legislation, a Citizens' Committee, which included some businessmen, was organized which may have helped somewhat in its final consideration. In any event, as stated before, this year in July Congress by legislation entitled "Congressional Budget and Impoundment Control Act" has established far reaching budgetary procedures which should certainly assist in controlling and rationalizing spending.

fifty-nine

But it is still a good question to ask how well equipped the business community is, as constituents of Congress, to help it develop what I will call appropriate congressional concepts of what to try to manage and what not to in the productive world of business in which so many live and work.

A Reexamination of the Nature of Congressional Exactments Relating to the Conduct of Business

When enacting a law intended by Congress to provide some remedial action what standards exist for guiding its action? For example, take pension reform, or legislation involving conservation, or pollution, or antitrust measures, or dozens of other laws. We know it seeks a remedial objective, but does Congress measure adequately the effect of the proposed remedy on operations in the business sector?

Many congressmen may attempt to poll their individual constituents to learn what they think about a matter as complicated as any one of these named subjects. That is unquestionably part of their job, however uninformative it may be on a complicated subject.

Polls are helpful, but congressmen are much more likely to get informed opinions on matters directly affecting their business constituents from those most affected by the proposed legislation.

To repeat, a very high percentage of people under retirement age are in groups we call businesses, and the remainder are mostly in groups which to a greater or lesser degree are dependent on business in the broadest sense. Why, then, should not the voices of business representatives be heard on vital subjects?

At present many businessmen do call on congressmen, giving their views on what they consider important matters. This, however, is an uphill climb and needs to be given recognition and legitimatized.

The Congressman in His Job

Would it be out of order to ask these few questions about our Congress, its organization, and its general performance?

Congressmen on the whole are well qualified as individuals but the question remains how many of them qualify themselves as legislators. And is there any post more important and more

demanding than serving the people in high posts of government—especially as legislators?

How is the average new congressman to learn about the
economic consequences of what he, as a legislator, does? How
can he fairly and expertly examine the probable course of action
his vote may bring about and its overall effect on the economy?
How does he go about solving the seemingly inherent conflict
between good economics and good politics?

So far as I know, there is little time for the elected congressman to study the history of regulations, and what regulations
have been tried and found lacking in effectiveness, or tried and
found detrimental. How does he avoid an unwitting effect
which reduces productivity in business, to the harm of everyone, especially employees?

At the moment, the practice is for what we know as "on-the-
job training," in committees, on the floor, and in the cloak
rooms. To be effective, he must, among other things, gain some
understanding of the "ins" and "outs" of the business world,
and of the economic effects of certain types and amounts of
taxation. If he is broadly based by experience or otherwise, he
has some knowledge of the international scene, and of the
trading rivalry between nations, and of the effects the support
policies in other nations of their businesses can have on competition with businesses at home.

Because of his selection by the voters, he rather quickly must
become almost a superman. The first thing we ask this newest
recruit to do is to help make the toughest decisions of every
nature, including economic decisions. We ask him to fully
comprehend the effects of that which he does. And we ask the
same of the congressman who has been in office for one or two
terms.

There is, of course, some safety in the newer members following the lead of those members of Congress who are more
experienced. Yet the system calls for each member to exercise
his own informed judgment.

*A New Hoover Type Commission to Consider
Congressional Performance*

In our complex society, as it nears the end of the twentieth
century, with our extraordinary economic programs, our finan-

cial and tax considerations, our international and antitrust problems, what Congress does and how it does it becomes of striking importance.

On the subject of what can be done to better the situation, what about a broadly based commission to examine the committee structure and the current operations within Congress itself and with the Executive branch? The Bolling Committee with its report to the House of Representatives of March 19, 1974 is a much debated effort. Its recommendations seem to have only partial acceptance.

Many, of course, will recall the Hoover commissions. They performed a real service in their examination of the Executive branch of government. Would it be in order to suggest the possibility of a "Hoover Commission" for the examination of both branches of Congress? It would be a monumental task. Congress itself would have the last word; but perhaps an objective study would help with the self-examinations all institutions need periodically.

A September 1974 Interim Report of the Joint Economic Committee of Congress relating to reducing inflation and restoring economic growth recognizes the situation, stating in part:

> The inflation of 1973-74 has given us new evidence that the Federal Government is not properly organized to deal with an extremely complicated economy. The private sector is today about twice as large as it was 20 years ago and infinitely more complicated. The public sector has also grown and now profoundly influences the private sector through its subsidy programs, procurement and stockpiling policies, credit programs, and regulatory activities. Furthermore, the U. S. economy is now more closely linked to the world economy as a result of the growing internationalization of production and global competition for scarce resources.
>
> Yet this changed environment in which U. S. economic policy must operate has not been matched by changes in the ability of the Federal Government to formulate policies to deal with it.[3]

While this report seems to place emphasis on the coordination of economic policy in the Executive branch, it does refer

---

[3] "An Action Program to Reduce Inflation and Restore Economic Growth," Interim Report of the Joint Economic Committee, Congress of the United States, Sept. 21, 1974. U. S. Government Printing Office, p 31.

to the federal government of which Congress is so vital a part.

I would not expect such a commission to concern itself with what I call the politics factor every member of Congress necessarily has in mind in decision making, a matter we discussed earlier. The best economic plans available can suffer complete distortions, defeat or lingering suffocation because of the politics factor. But there it is. That is the nature of the process, and I presume the country must and probably should live with it.

But the presence of this factor is not to say such a commission would not serve a most useful purpose.

The Regulatory Agencies

While we are about it, why not, in this era of increasing regulation, have the same commission or a different one take a look at the regulatory agencies? Some are relatively new and some long established. As indicated earlier, rightly or wrongly, new circumstances arise and new demands for regulations are made. Congress does what comes naturally, and, after adequate or inadequate consideration, passes a law. Many times the law provides for continuous regulation by a governmental agency. With Congress, as it must, going about its other business while retaining some supervisory control, a new regulatory commission comes into existence, a budget is developed for it, and its regulatory role commences with fanfare.

Recently, the Administrative Procedure Act was adopted which requires regulatory agencies in the exercise of their powers to provide due process—such as advance notice of policy actions and opportunity for the regulated and others to be heard, and the requirement that consideration be given to the views expressed at subject hearings. This due process requirement is one measure not to be overlooked or underrated. It may be one of the best means yet devised to protect against unworkable or improper regulation.

The regulatees' views regarding the necessity for particular regulations, the harm as well as the benefit the regulation is likely to cause, may suffer from imbalance but they should be heard. The regulatee needs a chance to point out the cost of compliance in relation to the benefit, the facility or product changes that will be required, and the duplication with other regulatory agencies.

So why not check the regulatory agencies searchingly and impartially? A reexamination of each agency's place in whole regulatory network and its performance should help the agency itself provided it checks out with merit. Some speakers and some reports at the recent summit meeting on inflation raised questions regarding regulatory agencies which need answering.

The President, in his nation-wide address of October 8, said, in part, "At the Conference on Inflation, we found, I would say, very broad agreement that the Federal Government imposes too many hidden and too many inflationary costs on our economy." Among other actions taken, he said, "I ask the Congress to establish a National Commission on Regulatory Reform to undertake a long overdue total reexamination of the independent regulatory agencies."

Decentralized Decision Making

Speaking about matters which business must continuously discuss with Congress, the regulatory agencies and any new Commission on Regulatory Reform which may be set up by Congress, I regard the desirability of decentralized decision making as of the highest importance. Who makes the decisions affecting production and where they are made in the operating process will largely determine the efficacy of corporate and other group efforts.

One thing is certain. Congress cannot run a business and get involved in the myriads of decision makings that take place daily. It cannot run one major business, let alone two hundred thousand or a million. With five hundred thirty-five members, that body is simply not organizationally structured to manage businesses. It is true, as I choose to put it, that Congress, through its laws, exercises a type of "management" function, many times through its committee system; and it is the attempt to exercise that "management" function which most concerns me.

It is, therefore, not only the general thrust of legislation which should concern business, but as important are congressional efforts at detailed management of what must be essentially decentralized operations if they are to function successfully. The same thing can be said of the regulatory agencies. Witness the rapid growth of the volume of regulations and manpower involvement with each month of the recent wage and price controls. Or compare the complexity of the innumerable sections of the Internal Revenue Code as it applies to business. Certainly one of the most useful reminders to any Congress is to exercise care that any new law preserves freedom of operation for business and does not smother it with red tape. After all, the objective is the general welfare of the people, not the building of regulatory agencies or the hindering of operations.

"Management" by Legislation

In this trilogy I have avoided trying to discuss "free enterprise," which rarely exists anyway. Rather, I prefer the pragmatic approach of discussing the real world with all its imperfections and its regulatory directions. We speak of the market system, and few understand we are actually comparing a system which fosters innumerable decentralized decisions with one which centralizes decisions. That is the heart of the market system. Its preservation and adequate freedom of operation may be desired for ideological reasons or desired because it works better and more equitably. I prefer to rest the case on the "works better and more equitably" base.

Without in any way equating the effectiveness of Marxism with our own system and whether he is creditable or incredible, I happen to agree with Aleksandr I. Solzhenitsyn in his evaluation of the USSR's extreme centralized decision making system, which incidentally appears to be moving toward more decentralization, when he says in his *Letter to the Soviet Leaders* published this year that:

> This Ideology that fell to us by inheritance is not only decrepit and hopelessly antiquated now; even during its best decades it was totally mistaken in its predictions and was never a science.
> A primitive, superficial economic theory, it declared that only the worker creates value and failed to take into account the contribution of either organizers, engineers, transportation or marketing systems. It was mistaken when it forecast that the proletariat would be endlessly oppressed and would never achieve anything in a bourgeois democracy—if only we could shower people with as much food, clothing and leisure as they have gained under capitalism! It missed the point when it asserted that the prosperity of the European countries depended on their colonies—it was only after they

had shaken the colonies off that they began to accomplish their 'economic miracles.' It was mistaken through and through in its prediction that socialists could never come to power except through an armed uprising. It miscalculated in thinking that the first uprisings would take place in the advanced industrial countries—quite the reverse. And the picture of how the whole world would rapidly be overtaken by revolutions and how states would soon wither away was sheer delusion, sheer ignorance of human nature. And as for wars being characteristic of capitalism alone and coming to an end when capitalism did—we have already witnessed the longest war of the twentieth century so far, and it was not capitalism that rejected negotiations and a truce for fifteen to twenty years; and God forbid that we should witness the bloodiest and most brutal of all mankind's wars—a war between two Communist superpowers. Then there was nationalism, which this theory also buried in 1848 as a 'survival'—but find a stronger force in the world today! And it's the same with many other things too boring to list.

Marxism is not only not accurate, is not only not a science, has not only failed to predict a *single event* in terms of figures, quantities, time-scales or locations (something that electronic computers today do with laughable ease in the course of social forecasting, although never with the help of Marxism)—it absolutely astounds one by the economic and mechanistic crudity of its attempts to explain that most subtle of creatures, the human being, and that even more complex synthesis of millions of people, society. Only the cupidity of some, the blindness of others and a craving for *faith* on the part of still others can serve to explain this grim jest of the twentieth century: how can such a discredited and bankrupt doctrine still have so many followers in the West! In *our* country are left the fewest of all! We who have had a taste of it are only pretending willy-nilly . . . .[4]

One can also agree with Dr. Arthur F. Burns in his address, "The Relevance of Adam Smith to Today's Problems," given in June last year at the Adam Smith Symposium in Scotland, when he said,

> Where free enterprise has flourished, nations have prospered and standards of living have risen—often dramatically. Where detailed governmental regulation has repressed individual initiative and

---

[4.] From pp. 41-43 LETTER TO THE SOVIET LEADERS by Aleksandr I. Solzhenitsyn. Translated from the Russian by Hilary Sternberg. Copyright © 1974 by Aleksandr I. Solzhenitsyn. English translation copyright © 1974 by Writers and Scholars International Ltd. Reprinted by permission of Harper & Row, Publishers, Inc.

stifled competition, economic growth has been hampered and the well being of the people has generally suffered.

But agreeing with both Solzhenitsyn and Burns does not alter our present situation, or should we call it our precarious predicament.

We are a part of a regulated society. Those regulations are, in my view, certain to increase. Talking about capitalism or free enterprise or socialism will not make them go away. It is a very pragmatic world in which we live amidst all its warts and its shortcomings.

We have labor and management, and some form of regulation is needed there. We have held over war powers which we could well revise or perhaps do without. We have the poor, the old and the ill who for one reason or another cannot cope. We know the inadequacy with which controls have functioned as far back as the Greek and Roman experience and, personally, I think it a good thing this is so. But we also have investors who will be accorded some form of protection. We have customers who survive in a credit structure. We have large, small, and medium sized corporations which give employment, produce desirable products, and perform useful services but which sometimes create undesirable waves. For these and many other reasons we are essentially a regulated society.

Restrained anger at those who do not appreciate the usefulness of freedom in group operation has little usefulness. Our problem is how much government management of business will we have and what kind will it be. How decentralized will the decision making be? As a goal, as much freedom as possible, but in the age of nuclear power how much freedom is actually desirable in the proliferation of nuclear energy?

Instead of turning off some of our listeners by extolling free enterprise as a concept, as Marxism turns us off, it seems to me we would get further in a world of growing regulation by discussing the pragmatic value of the decentralized approach to group operations.

To repeat, when it comes to working with Congress it seems to me that stressing decentralized decision making because of the functional results is a good way to emphasize its benefits. And, as I have said, it is about the most important thing a congressman can keep his eye on when considering the merits or demerits of legislation.

It is very human to want to manage. This applies to Congress in its relation with the Executive branch of government. It applies to both when it comes to prescribing how business must be conducted. It is in the area of attempting to assert management control by government that it can be said with a will that, for every conceivable reason, that government is best which governs least.

Lately there have come from the academic world serious proposals for removing barriers to expected market reaction, especially in the competitive area and downside pricing. To this end it is suggested that agriculture restrictions on interstate movements be removed and production quotas and export subsidies be reexamined; that transportation routes and rates become less restricted and more competitive, and subsidies for shipping be abolished; that natural gas control at the well-head be ended; that petroleum allocations and oil price controls be terminated; that banking freedoms relating to interest payments and areas of competition be enlarged; that resale price maintenance be ended; that the antitrust laws be revised; that merger guidelines be reformulated; that changes be made in labor laws relating to membership, apprenticeship, hiring halls, and government contracting; that limitations on imports be modernized; and that some government operations be made subject to competition.

Doing something about all this would be an arduous task but is it not timely to make the attempt?

Another serious matter in which business can be helpful to government is in the area of the formation of capital, its sources and its uses. Briefly, it is all part of the decentralized business operational picture. New capacities, whether of a basic nature as power supply or steel capacity, or such things as a new shopping center, or the creation of a consumer service, all need capital in greater or lesser quantities and all are of direct benefit to consumers.

The nation now faces a capital shortage, a growing one. This, many believe, is due primarily to the lack of appreciation of the effects by Congress of the legislation it adopts, whether in the field of taxes, of social benefits or other legislation. Without developing this highly important subject further, let it be said that businessmen are peculiarly fitted to discuss this matter with facts and conviction and some have. They see the uses and the vacuum better than most. Congress should be willing, and I believe it will be, to listen to properly presented statements on this matter.

In general, Congress need have no apprehension of greater Washington efforts by business. It has plenty of authority and certainly enough will to resist or restrain nonuseful proposals. It has the last word always. It makes the final decisions. If business does a constructive job, that end result will be a better one for everyone.

Conclusions

May we retrace our steps for a few moments to observe some of the entries in this three-part story.

In the beginning we analyzed the gradual and, as in the recent case of the oil industry, at times precipitant embrace by Washington of business decision making. This paper concludes that with the growth of population, rising aspirations of people, new forms of transportation and communication, enlarged international relations, modern inventions of great variety, urbanization, conservation, ecology, changes in modes of living, specialization in work, and the general growth of interdependence of people within the confines of our nation, as well as internationally, and with more governments abroad taking over raw material supplies and ownership or management of basic producers, plus an array of other mounting factors, it becomes virtually inevitable that the Washington embrace will continue and tighten.

This paper also concludes that it is high time for business to terminate a standoffish resentment toward Washington and its interference in business affairs. Rather, we considered the alternative and pragmatic direction of business moving toward a more cooperative attitude with government, a shared responsibility type of approach, with better communications between the sharers of the responsibility.

We also discussed the advantages of persuading wave after wave of newly elected members of Congress on the Hill and their appointed staffs to think in terms of the effect on the nation's production and service businesses of all kinds and sizes, of the various proposals for legislation.

In part two of this series, we discussed the seemingly eternal conflict between good economics and good politics. A brief review of several specific cases was undertaken, including the President Kennedy—steel industry 1962 price incident.

Given the premises upon which this paper is predicated, of a new attitude on the part of business and a somewhat receptive government, better long range economics should have a chance in the future in competition with what may seem like good short range politics at the time. The recent summit meetings on inflation seemed to be a valid step in this direction.

It was concluded, in fact, that good economics must necessarily be good politics over any reasonable span of time.

We also considered the various relations of Congress and business, and the vicissitudes of learning to live constructively with a governing body such as Congress with its constant changes of personnel and attitudes.

We even considered, at the risk of possible impertinence, of suggesting that a new examining look be taken at Congress, its committee structure, its attempts at business "management" and how it functions in the public interest. Gingerly, and with trepidation, the suggestion was offered that, along with the rest of the country generally, business, representing in part as it does through employment and stockholder interest, the great majority of the working and investing population of the nation, had an interest in scoring congressional performance, even to the point of suggesting a replay for Congress of the type of examination done in the fifties by the Hoover commissions.

There is here and in our discussions much, much more. But I cannot end these interludes without a few observations.

Much as we need to live in today's world and its troublesome dilemmas, I hope we will not be content with that. In the year 2,000 and later, people will need their daily bread just as they need it today. An ability to live in a more closely governed world with a modicum of cheerfulness will stand all of us in good stead.

For explained reasons I find much that is here addressed to business. Collectively, it is composed of groups of some of the finest manpower our nation possesses. It is at least the equal of other groups here or abroad. It has the capacity to move with flexibility from task to task. Some from business, no doubt, will from time to time join the ranks of those who govern the rest of us. There is real satisfaction of the spirit, whether in or out of business or government, in making your contribution.

In this troublesome era it is well to remember that our grandchildren and great-grandchildren and their offspring will do more than we ever dreamed of doing. They will climb higher mountains than we have climbed, many we never knew existed. They will look back on what we have attempted and partially accomplished. I hope they will find it "good" in the broadest sense of the word.

PRESIDENT KENNEDY AND STEEL PRICES

by

ROGER M. BLOUGH
Retired Chairman of the Board
United States Steel Corporation

Prepared for the
JOHN F. KENNEDY LIBRARY

NATIONAL ARCHIVES AND RECORDS SERVICE

of the

GENERAL SERVICES ADMINISTRATION

380 Trapelo Road

Waltham, Massachusetts 02154

February 1975

Table of Contents

I   President Kennedy and Steel Prices 75

II   Early Meetings with Kennedy 76

III The 1962 Steel Negotiations 78

IV Seeking a Price Increase... 81

V   April 10, 1962... 84

VI Labor Negotiations and Price Increases 85

VII The Volcanic Eruption.. 88

VIII Planned White House Intervention in Wage-
      Price Action.. 90

IX The Tidal Wave of Official Pressure 94

X   Steel Prices Lowered in the Storm 101

XI The Public Reaction 103

XII The Media—And My Response to President Kennedy.105

XIII Tuesday, April 17, at the White House 107

XIV Aftermath...110

XV Reappraisal..113

I. President Kennedy and Steel Prices

On April 11, 1962 President John F. Kennedy launched a harsh and unprecedented attack on United States Steel Corporation and other leading steel makers, denouncing an increase in steel prices as the action of "a tiny handful of steel executives whose pursuit of private power and profit exceeds their sense of public responsibility" and who show "utter contempt for the interest of 185,000,000 Americans."[1]

In his press conference on that date he called upon the Department of Justice and the Federal Trade Commission to examine the price action. He asked the Department of Defense and other agencies to review their procurement policies. He told of plans, undoubtedly inspired, among members of Congress to make inquiry as to how steel price decisions were made.

The President's news conference, together with other actions here related, had the effect of generating a political furor in Washington, with Congressional members of both parties and numerous units of the Executive branch of government caught up in it. It rippled out across the country borne by headlines and newscasts. It brought forth public statements and editorials, investigations, and proposals, and continued long after the steel companies had rolled back their prices to their former levels.

A wage increase in steel had triggered the price increase and many events had preceded and followed the President's statement. At the time of these events I was chairman of the board and chief executive officer of United States Steel continuing in that office until retirement in January 1969. Thirteen years after the episode and after a number of requests, including requests from the John F. Kennedy Library, it seems that a personal recounting of the Kennedy steel price story can be of interest and useful to those concerned with government and corporate relations and the future of the market system.

---

1. Exhibit A—*The New York Times*, April 12, 1962, p. 20. (Portions of conference relating to steel.) *See page 117.*

II. Early Meetings with Kennedy

My first meeting with John F. Kennedy, arranged by Hal Korda, was early on a summer's day in 1960. Mr. Kennedy was then a senator from Massachusetts. Hal Korda was in the automobile leasing business and, as I recall it, had a long time personal relationship with Senator Kennedy.

A few businessmen were invited to the apartment of Henry Alexander, chairman of Morgan Guaranty Trust Company. Senator Kennedy was running for the Presidency and introductions to businessmen were a part of a "getting to know you better" program. The views of Senator Kennedy on a number of issues of interest to business were stated at this meeting and I believe he made a good impression.

Earlier, during the 1959-60 steel strike which lasted 116 days, I had received a call from Senator Kennedy who was then chairman of the Labor Subcommittee of the Committee on Labor and Public Welfare. He indicated he wished to be helpful in any way he could in ending the strike. He said nothing about his relations with David McDonald, the president of the United Steelworkers of America.

I thanked him for his interest and said that while I saw no opportunity at the moment for him to be helpful, his interest would be kept in mind.

My recollection is that the call came during the period of several weeks when Richard Nixon, then Vice President, was conferring separately with David McDonald and then with me at the Vice President's home in an effort to facilitate settlement.

After Kennedy's election as President in November 1960, the first occasion to meet with him arose out of the correspondence which he initiated in September 1961 against a background of a rumored advance in steel prices.

The President's letter of September 6, 1961,[2] individually addressed to chief executive officers of twelve steel companies, referred to the history of steel wages and prices since 1958 and to another impending wage increase under the 1960 contract. The letter asserted that even after the October 1 wage increase took effect, the steel industry could "look forward to good profits without an increase in prices." This forecast was simply

2. Exhibit B. *See page 121.*

wide of the mark. He also asked that the industry forego a price increase "now" so it could enter the labor negotiations "next spring" with a record of "three and a half years of price stability."

In my response of September 13, 1961[3] I said "we in United States Steel cannot forecast the future trend of prices in any segment of the steel industry and have no definite conclusions regarding our own course during the foreseeable future,..."

I pointed out "that from 1940 through 1960 steel prices rose 174 per cent, but the industry's hourly employment costs rose 322 per cent, or nearly twice as much."

Included was a quotation from the report prepared for the Department of Labor by Professor Livernash of Harvard University:

> Obviously while price policy can be debated in the short run, in the long run all cost increases must be met. Steel has done no more than this.

My letter also referred to the theory that steel is a price bellwether pointing out

> there was no increase whatever in steel prices during the years 1940 through 1944; but this did not prevent a substantial inflation in wholesale prices generally during the same period. Conversely from 1951 to 1956 there was virtually no net change in the wholesale price level, despite the fact that steel prices advanced by about 30 per cent.

I said that I was glad to see in the President's letter reference to the urgency of preventing inflationary movements in steel wages and that "by absorbing increases in employment costs since 1958, [the steel industry] has demonstrated a will to halt the price-wage spiral in steel."

Soon after this letter was mailed, I received a telephone invitation to come to the White House to see President Kennedy. During this visit on September 21, 1961, the President said he doubted the usefulness of further letter writing. He in substance asked if we had any plans with respect to price changes. I advised that our management had had further discussions since my letter of September 13 and had concluded that market fac-

---

3. Exhibit C—Letter to President Kennedy, September 13, 1961. *See page 123.*

seventy-seven

tors did not warrant an immediate attempt to change prices especially in light of the upcoming wage negotiations.

I again pointed out our cost increases and while not stating definitely that we would increase prices after the next wage negotiations, because no decision had been made, clearly indicated to him the necessity for reviewing the price situation at that time.

This September 21 meeting was cordial enough. Early in October Hal Korda called a United States Steel executive whom he knew with this message which, as relayed to me, I quote only for the sake of completeness:

> The President was very much impressed and pleased with Mr. Blough. The President had asked Korda to see that in as many ways as possible this is conveyed to Mr. Blough.

The President's impression appeared to undergo a radical revision on April 10, 1962.

III. The 1962 Steel Negotiations

On January 23, 1962, at the initiative of President Kennedy, David McDonald and I met with him and Arthur Goldberg, Secretary of Labor, to discuss the impending steel negotiations. Mr. Goldberg had been general counsel of the Steelworkers Union for a number of years prior to being named Secretary of Labor and was quite knowledgeable in labor relations.

President Kennedy expressed the domestic and international importance of our negotiations. He spoke of a balanced budget in the same context. He recognized the unfavorable results to steel producers and employees of the build-up, in anticipation of a strike, of steel inventories in the hand of customers, to be followed after the negotiation by a period of very low steel manufacturing operations and layoffs while customers reduced their inventories. He hoped for an early settlement and one which would not involve a price change.

There were statements by Arthur Goldberg and David McDonald bearing on the negotiations but if there was any suggestion of a "social contract," which I will discuss later, in the case of steel it escaped me completely. Incidentally, David McDonald later indicated he agreed to a 3 per cent ceiling on the increase in the impending negotiation. This also is not part of my recollection and it is difficult to reconcile with his concluding a labor package at a 2.5 per cent cost, a cost lower than one for which he had Presidential sanction.

For my part, I reaffirmed my previous position that wage negotiations and price negotiations were separate matters, that I did not wish to talk about price matters with the union. The others present understood, I was sure, the legal and practical reasons for this.

It was made clear that while I appreciated the "public interest" in the matter, in view of the costs previously incurred since 1958 without price changes, I did not want anyone present to be under any misapprehension of what effect additional costs might have in the industry.

I pointed out there had been no dividend increases for five years, that United States Steel's plant and facility expenditures were required to be restricted, that many projects were held back for lack of money, projects which would make us more competitive with Europeans and thereby provide more American employment. I said that there were not enough earnings after paying expenses to have a strong steel industry and that corporate managers had to think of all phases of the business not just labor problems.

I mention this in some detail because of the assertion by some that President Kennedy had been double-crossed when United States Steel attempted to raise steel prices following the negotiations. He is reported to have said something to this effect to David McDonald when our prices were raised. Nothing could be further from the truth. If he was double-crossed he must have double-crossed himself. Perhaps he meant that his hopes for price stability had not been realized. Or perhaps he expected that his communication of a wished-for price result would be accepted as a command performance.

As I explained earlier, President Kennedy neither asked for nor was given by me or anyone else in United States Steel a commitment, implicit or otherwise, with respect to steel prices in connection with the 1962 steel wage negotiations, regardless of what some who wrote about the incident inferred.

President Kennedy, himself, confirmed this during his press conference on April 11, 1962, one day after our price increase.

Did you—is the position of the Administration that it believed it

seventy-nine

had the assurance of the steel industry at the time of the recent labor agreement that it would not increase prices?

A.—We did not ask either side to give us any assurance, because there is a very proper limitation to the power of the Government in this free economy.

* * *

We never at any time asked for a commitment . . . because, in our opinion, that it—would be passing over the line of propriety.

But I don't think that there was any question that our great interest in attempting to secure the kind of settlement that was finally secured was to maintain price stability, ...[4]

President Kennedy's response to the reporter's question indicated that he concurred in the position I had earlier asserted that prices were not a proper subject for discussion in this labor negotiation.

I doubt very much if United States Steel, under the circumstances then prevailing, would have entered into a costly wage negotiation or agreed to any settlement if our freedom and flexibility of pricing had been involuntarily taken from us before the negotiation began.

United States Steel had found it necessary to raise its prices in 1946 following the steel wage settlement. The same was true in 1947 and 1948. It was also true in 1950, in 1952, in 1953, in 1955, in 1956, and in 1957. There was no other alternative in view of rising labor costs and a relatively low improvement in output per man hour.

Direct wage and salary costs represented over 40 per cent of all costs. During the period from 1946 to 1960 steelworker employment costs had risen at a rate of about 8 per cent annually while during the same period the best we could do to improve productivity was about 2 per cent.

The difference of more than 5 per cent cost increase annually had to come from those who purchased steel in the form of price increases.

Our business economic problems and his analysis of our financial situation were not, however, the basic cause, I believe, of President Kennedy's violent reaction. Following the long strike in 1959 and early 1960, the market situation and prior public statements made during the negotiation were such that

4. Exhibit A—Kennedy Conference.

in United States Steel we concluded not to attempt a price increase. In economic terms this resulted in United States Steel absorbing higher costs and in holding its prices relatively stable for what turned out to be about four years, from 1958 to 1962. Richard Nixon, then Vice President, had been helpful in settling the 116-day strike. Inadvertently our price restraint after the 1960 negotiation laid the foundation for President Kennedy and some of his associates to claim that we deliberately followed a pricing policy which favored Richard Nixon and discriminated against the President. Looked at politically, the labor settlement assisted by Vice President Nixon in early 1960 was not followed by a price increase. (Here again there was no discussion with the parties nor any commitment given regarding prices.) But the negotiations in 1962 in which President Kennedy had played a part were followed by a price increase. This could be construed as a public image setback for President Kennedy and undoubtedly was so construed by him.

Certainly the President's desire to contain inflationary influences was an important factor as his 1961 letter to me indicated. But other facets which built up his expectation and hence his disappointment of which we were not aware and which are discussed later, plus the jarring of his political image, were basic to his rather unrestrained reaction.

When a reporter asked at the President's press conference on April 11 whether "there was an element of political discrimination" in our increasing price, President Kennedy was visibly disturbed by the question and, as a reading of his answer indicates,[5] hesitantly replied in an answer which includes this sentence: "If they do it in order to spite me it really isn't so important."

IV. Seeking A Price Increase

With the major aspects of the labor negotiations concluded and announced on March 31, 1962, the management of United States Steel turned its attention to the omnipresent price problems and the inadequacies of earnings.

We had given much attention to this matter in the prior

5. Exhibit A—Kennedy Conference.

weeks. The factual support for a price change had been assembled and was part of the deliberations of the top operations committee, when it took the unusual step of meeting on a Saturday, April 7. The meeting once more reviewed all aspects of our cost price situation. It was concluded that we would recommend to the regularly scheduled meeting of the Executive Committee of United States Steel's Board of Directors on April 10, that a price increase of 3.5 per cent be made on steel products. This averaged about three-tenths of one cent a pound, or $6 per ton. A draft announcement was reviewed, and finally agreed upon.[6]

At the Executive Committee meeting on the 10th, a Tuesday, after full consideration of the proposed increase, approval was given. The proposal was discussed with some of the other directors who concurred in it.

Price increases, notably those in steel, are never done at the right time, in the right manner or in the right amount, according to the critics. Of all the inviting occasions for Monday morning quarterbacks, a major price change in an industry such as steel affords one of the finest opportunities. This particular change, it was said, was not timely. It should not have been made "across-the-board" but should have been selective. It did not take into account the effect it would have on the imports of foreign-made steel. The economic climate was not right. There was also, as previously noted, the thought that the reaction of the government had not been properly assessed.

Changing prices is largely a matter of competitive judgment, after demand is assessed and costs and other factors including domestic and foreign competition have been weighed. In this case, it was the commercial judgment that under the necessities of the situation a 3.5 per cent increase applied to all products was the best and most equitable change to attempt. We knew it was too much to expect that the steel consuming markets would welcome the higher cost of steel. Purchasers in many cases were also feeling the pinch on their profits.

Our cost increases had been mainly "across-the-board," that is, they applied to all products. They consisted of increases notably in labor costs, in taxes, in transportation expenses, and in other costs which were generally applicable. As to the

6. Exhibit D—United States Steel Press Release dated April 10, 1962. *See page 128.*

amount, since we had been without price relief from 1958 until 1962 and had suffered a narrowing of profit margins, the amount finally selected was low, if anything.

Management has an inescapable responsibility to do those things necessary to keep a company in financial health, and to maintain it as a flourishing job-providing element of the American economy.

In this case, there was a certain amount of improvement in market demand which we thought would help to sustain the price change. In fact, as the market developed subsequently to April 10, there would probably have been no time in the months which followed more commercially opportune to announce a general price raise.

We had, if anything, delayed overlong in raising prices, although there were reasons for this.

Going back to 1959, in connection with the steel labor negotiations of that year, United States Steel made this pledge on July 28:

> Whatever the length of the strike, and whatever the eventual outcome of the negotiations—so long as they are voluntary—we in United States Steel do not intend to raise the general level of our prices in the foreseeable future.

It will be recalled that in that negotiation we were dealing with an employment cost increase that had averaged almost 8 per cent annually since 1940, and this statement was expected to assist in obtaining a less costly contract. When the 116-day strike finally ended in early January of 1960, this pledge plus market conditions seemed to rule out, we thought, any reasonable chance of making a price increase that would hold in the marketplace.

Possibly, a price increase could have been made at the time of the wage increase, October 1961, but, generally speaking, we did not favor price increases just before or during labor negotiations. The need for a price increase had been starkly apparent for some time, in spite of what President Kennedy's advisers had said in his letters to steel producers of September 6, 1961. It is my surmise that if market conditions had permitted, and we had not been on the verge of active labor negotiations, an attempt to raise prices might have been made by our company or by a competitor several months earlier.

eighty-three

V. April 10, 1962

April 10, 1962 was a busy and momentous day.

After a decision is made to raise prices, it is desirable to make the change as quickly as possible. For this reason, when the Executive Committee had acted, I promptly telephoned President Kennedy's office requesting a few minutes of his time that day, if possible. His secretary, Mrs. Lincoln, confirmed the arrangement, and about 5:40 P.M. I arrived at his office in the White House.

Tuesday, April 10, was a fairly warm day, with a temperature in Washington of 67°. Twenty-four hours later, the official temperature was to fall 20° but the political temperature in that city was much hotter than it had been in quite some time.

Calling on the President and discussing our price increase with him personally was, I thought, the courteous thing to do. It would have been alien to our concept of government corporate relations and prerogatives to have asked his permission in the absence of legislative authority. The alternatives to seeing the President would have been to send him the announcement or let him read the news in the press. I thought the proper thing was to see him personally.

Upon meeting the President, I informed him of United States Steel's decision to raise the general level of its steel prices for the first time in four years and gave the reasons for that decision. I also handed him a copy of the proposed press release which he read. Soon after 6:00 P.M. the news release was made from the New York and Pittsburgh offices of United States Steel and the first news appeared publicly at 7:00 P.M. in the form of an Associated Press bulletin to its member papers.

President Kennedy was visibly annoyed by the fact of increasing steel prices. He immediately called for Secretary of Labor Arthur Goldberg, who came in a very short time to the President's office and read the proposed press release. There ensued a rather acrimonious discussion, mostly on the part of Arthur Goldberg and myself.

I left with the impression that President Kennedy and Mr. Goldberg were apprehensive for reasons not clear to me and from a background I did not possess. This seemed to be related to organized labor in some way and since I had a fair understanding of expected Steelworker Union reaction, the problem must lay elsewhere. Later we learned of some of the circumstances behind this concern.

In United States Steel we, of course, knew of President Kennedy's desire to stabilize prices and, in a way, in his "price stability" statement following the wage negotiation, he had gone out on a "limb" which I am sure he knew. But we did not anticipate the storm that broke. Everyone connected with the steel industry was well aware of the many governmental efforts that had been made, particularly at the time of steel labor negotiations, to prevent steel price increases, mainly by utilizing the force of public opinion. What we did not know about were all the background efforts of Walter Rostow, Arthur Goldberg, Walter Heller, and President Kennedy, with respect to the automobile negotiation and Walter Reuther, and the earlier orchestration of Senate opposition in steel.

There was also at the time some speculative comments about a desire on the part of United States Steel to challenge the President or the Presidency. Of all the imaginable uninformed comments this seemed to us to be the most absurd. Naturally we wanted a price increase. After four years of rising costs and no price increases, business judgment and competitive conditions dictated the action taken, not any desire to hurt the Presidency or to assume a political or partisan role.

Some time during the day after the price announcement by United States Steel, or the following day, similar announcements were made by Bethlehem Steel Corporation, Republic Steel Corporation, Jones & Laughlin Steel Corporation, Wheeling Steel Corporation, and Youngstown Sheet & Tube Company. These prices generally followed those made by United States Steel. But the drama was only beginning.

VI. Labor Negotiations and Price Increases

For a time the question persisted as to why United States Steel did not either raise its prices during labor negotiations or state more specifically what it would do on prices when the labor contract was concluded.

There are a number of reasons why in the circumstances that prevailed in 1962 it would have been ill-advised, we thought, to be more explicit about price increases than we were.

There is the question of alleged violation of the antitrust laws. For example, after a 3 per cent increase in steel prices went into effect on August 1, 1958, shortly after a 5½ per cent increase in steel labor costs, Senator Estes Kefauver (Democrat, Tennessee) conducted hearings on that increase before his Judiciary Subcommittee on Antitrust and Monopoly.

It was brought out there that reporters had questioned a number of steel industry executives representing different companies and they had said there should be a price increase and that one was necessary, although no one said what he intended to do.

Senator Kefauver elicited from Victor R. Hansen, then Assistant Attorney General in charge of the Antitrust Division, what Senator Kefauver as one of the supporters of Senator Gore in the oratorical barrage in the Senate against a steel price increase (August 22, 1961) described as Judge Hansen's "interesting suggestion that the steel companies might be communicating with each other as to a change in steel prices through newspapers and trade journals." Any obtrusion of prices into multicompany labor negotiations could easily have led to suspicion on the part of the Justice Department.

Another reason for holding final price intentions confidential is to retain the freedom of action to accommodate the market changes, competitive changes and similar circumstances. In many cases you are somewhere in a three or four months time frame and you do not know where you are time wise. The results of labor negotiations, moreover, influence price decisions and it was not reasonable under circumstances of that day to say what your decision would be until you knew what your labor contract results were.

My own view was that announcing a price rise in advance of negotiations could possibly have a negative effect on a reasonable result in those negotiations.

Six weeks after my reply to President Kennedy's letter, and during the negotiations, I was questioned at a news conference on October 31, 1961 about prices and said

> If you are asking me at the moment what I think about the profit situation in the steel industry, it is far too low. As I said a number of times, if you are asking me to forecast anything with respect to

prices I can do nothing other than repeat what I said with reference to it in the answer to President Kennedy.

At the January 30, 1962 news conference, I made similar remarks pointing out that United States Steel's earnings statement demonstrated what we have been saying all along: "that costs have been going up in a highly competitive industry when there has been no movement in prices." I pointed out that the cost situation had deteriorated to the extent of about 6 per cent since 1958, the last general steel price change, and that one of the major problems that lay ahead for the steel industry is the proper correction of "the maladjustment that exists." I went on to say that, "the cost-price relationship has to be adjusted moderately in favor of the kind of thing that will produce jobs, which is more profits."

The next day *The Journal of Commerce* headlined a story in these words: "Big Steel Eyes Lower Costs, Higher Prices," which certainly should have been ample notice to President Kennedy and his staff if any notice was needed. This latter press conference was held after our meeting with the President in January.

On February 26, 1962, in an interview conducted by *U. S. News & World Report*, after pointing out that during the three preceding years hourly employment costs in steel had gone up at more than twice the rate of increase in shipments per man hour, I said,

> And you're asking me how long that can continue to increase and how long it can be borne without some kind of remedy? I would give you the answer that it's not reasonable to think of it as continuing. In other words, even now there should be a remedy. If any additional cost occurs, the necessity for the remedy becomes even greater.

Would that have been said if there was a "social contract" or any understanding regarding price?

I recall the discussion several of us in United States Steel had immediately following the Kennedy call to Conrad Cooper regarding the settlement, in which, among other things, he said, "It is obviously non-inflationary and should provide a solid base for continued price stability." The wording of his message appeared to contemplate no price increase and this was a problem, since we were considering recommending one to the directors. Conrad Cooper could not have been expected to reply to the President on prices. While a member of the top management committee of United States Steel, his responsibility was labor relations and not prices.

VII. The Volcanic Eruption

Hal Korda played an interesting and important role in the bizarre events which followed. My recollection is that he was in the White House for some reason on the evening of April 10, the day prices were raised, perhaps called there by the President because of his rapport with business. He telephoned William Lang, an officer of United States Steel whom Hal knew, at his home well after midnight and asked, "What is being done on the price problem?" He reported that the "lights were on at the White House" and "The Man" was waiting for some immediate action or commitment. There were references to possible actions the White House was considering if an immediate rollback was not forthcoming. The atmosphere portrayed by Korda was that of a volcanic eruption.

Mr. Lang told Mr. Korda that nothing could be done at that hour, that early on April 11 others in management would be informed of the President's request and consideration would be given to the questions raised. Next morning, Wednesday, Lang told us of the call. And thus began three days of intensive activity.

It is difficult to define Hal Korda's role. He was a man in business for himself, with many industrial, banking, and commercial contacts. Mr. Lang thought of him as having an appreciation of economics, the determinations of the marketplace, and competition. As an intimate in the White House and very alert to the political processes, with his understanding of the issues, he was in a unique position and apparently President Kennedy utilized his services in this price emergency.

The post-midnight call from Korda was followed by many others during the next three days conveying thoughts and trying to act as a conciliating party.

Several times he phoned me directly. He also gave us a number where we could call him in the White House.

I have no way, of course, of checking the information he gave us but at the time it sounded factual. He referred frequently to Robert Kennedy, the Attorney General, as a strong motivating influence in the many sided drive against steel. He spoke of Robert Kennedy ordering the FBI to call on a newsman at 3:00 A.M. to learn precisely what the chairman of Bethlehem Steel Corporation had said at a stockholders' meeting on April 10 in Delaware.

Korda also indicated the serious dispute that arose between J. Edgar Hoover and Robert Kennedy when Hoover learned the next day of Kennedy's use of the FBI.

Mostly Korda was acting for the President or his staff on his calls, which sometimes came to William Lang or Robert Tyson, chairman of the Finance Committee of United States Steel, or to Bradford B. Smith, chief economist.

It was mainly as a result of a Korda call that I went to see President Kennedy at the White House on Tuesday, April 17, one week after the price rise, although the thought was mentioned at the meeting in the Carlyle Hotel on April 14 which I will refer to later. I understood the call was made at the suggestion or with the concurrence of the President.

Korda led me to believe that Theordore Sorensen, then a member of the White House staff, had written the Kennedy attack on the steel industry delivered by the President on April 11. He also told how someone had dug up personal income tax returns of steel executives and their expense accounts. The purpose appeared to be to "get something" on the steel executives.

While my salary in 1962 was a good one, I thought my business expenses were modest. Korda quoted the President as saying my income tax deductions and my expenses were mere "peanuts" and there was nothing there that could be used against us.

Hal Korda also related, again I suppose at the instigation of the President, the considerable worry the President developed when he saw the cumulative results of Executive, Congressional, and press response to his attack on the steel industry. The President, in effect, confirmed this in my meeting with him on Tuesday, April 17, by expressing surprise and regret at the magnitude of public and business reaction and a wish to find some means of calming down the situation.

Although a relatively young man, and whatever his acquaintance with business economics, President Kennedy was very knowledgeable about history. From the standpoint of representative government he seemed to sense that his demonstrated ability to stir up the country could act as a two-edged sword, especially if a similar pattern were followed on some future occasion by one of his successors whose motives might not be as innocent as were President Kennedy's.

It was also from Hal Korda that I first was informed of the personal political reaction of President Kennedy, the loss of political "face" as it were. Korda mentioned several times the internal White House adverse reaction to the "different treatment" accorded Richard Nixon following steel negotiations and that given President Kennedy. The point seemed to be that United States Steel had the "power" to hand out political acclaim, or the reverse, at will. The economics of running a business seemed to be completely lost sight of in the heat of reaction to the price increase.

In passing it may be appropriate to point out this reaction as one of the unprofitable by-products that seem to be generated when politics get too much involved in marketplace decision making.

VIII. Planned White House Intervention in Wage-Price Action

Over the years it has become quite apparent that many things go on in and around the White House with respect to labor negotiations about which there is very little public knowledge. The Kennedy Administration in 1961 and 1962 was no exception. Only a part of the story was known to the steel negotiators. A major part was revealed later and undoubtedly accounts, at least in part, for the type of attack made by the government at the time of the steel price increase.

I am indebted for part of the story to Walt W. Rostow[7], then a Presidential Assistant, and for another part to Professor William J. Barber of Wesleyan University.[8]

Walt W. Rostow apparently urged the new President to press for a wage freeze and a price roll-back in key sectors of the

7. *The Defusion of Power 1957-1972* by W. W. Rostow. See Chapter 15.

8. Material prepared for a conference of experts on the Development of Wage-Price Policy in the United States, held in Boston, November 1 and 2, 1974 under the sponsorship of The Brookings Institution.

economy—with steel and automobile industries as the first targets. Apparently Mr. Rostow's view was then (as it still is) that wage-price "treaties" should be obtained in which labor would agree to forego increases in money wages on the understanding that productivity gains would be converted into price reductions. This would, in his view, permit expansion without inflation.

Rostow, on Presidential instruction, conveyed his proposals for wage-price "treaties" in key industries to Secretary of Labor Goldberg. Goldberg expressed some opposition to an immediate wage-price agreement for what he thought were good reasons: that the relationship of wage-price policy to the serious international competitive situation was not understood and that industry-labor agreement on wage-price policy was more complex than its proponents realized. Rostow persisted, believing that Walter Reuther, president of the United Automobile Workers, was sympathetic to such an arrangement.

Apparently there were marked differences of opinion among the Presidential advisers. One view was that government intervention tended to produce wage settlements higher than those when there was no intervention. There also appeared to be matters of strategy differences among those who wanted intervention, one group believing that in the absence of direct controls the policy should be policed either by an aroused public opinion or by unpublicized Presidential communications with leaders in business and organized labor. The other group believed in the use of productivity as a standard. It held, among other things, that in industries where the increase of productivity was less than the national average (which, incidentally, was true of steel), prices might have to increase, but to have price level stability it would be necessary to have reduction in some prices. Hence, productivity was introduced as a test of responsible price making.

It should be noted in connection with the treatment of the steel industry on prices in April 1962 that the Economic Report of the President, 1962, in discussing guideposts sanctioned appropriate increases in prices where "the level of profits was insufficient to attract the capital needed to finance a needed expansion in capacity . . ." a situation abundantly true in steel.

In June, Walt Rostow talked with Walter Reuther. After some

negotiations Reuther inferred that he could not agree to such an arrangement or "social contract" without knowing what Dave McDonald would do in steel negotiations. This reaction was a natural one in relation to union rivalry.

Arthur Goldberg, according to Walt Rostow, monitored the automobile settlement urging restraint in the labor negotiation, although the automobile industry apparently did not reduce its prices. The Rostow idea of a wage freeze and a price reduction did not materialize.

When it came to steel, the problems were even more complex.

After the automobile negotiations were well under way, what now appears to have been a formidable plan of attack on steel was organized. Professor Barber indicates that sympathetic senators under the leadership of Senator Albert Gore (Democrat, Tennessee) spoke in the Senate apparently using briefs prepared by someone in the Council of Economic Advisers. The theme of these inspired briefs dwelt on the importance of stable steel prices.

On August 22, 1961 Senator Gore, on the floor of the Senate, called on the President to exercise control over steel prices because Mr. Gore was convinced "that a plan is afoot for the steel companies, acting in concert, to raise the price of steel by an appreciable amount." On the Democratic side, Senators Douglas, Clark, Symington, Humphrey, McCarthy, Kefauver, Long, Young, Sparkman, Monrooney, McGee, Mrs. Neuberger, and Moss spoke in the floor debate in addition to Senator Gore and in support of his position.

There were rebuttals from Republican Senators Dirksen, Javits, Bush, Goldwater, Saltonstall, Bennett, Capehart, Scott, Kuchel, Hruska, Allott, Keating, and Miller. The legitimacy, among other things, of governmental intrusion into pricing decisions in the private sector was questioned.

This was the first time to the best of my knowledge that a President of the United States had been publicly called upon in the words of Senator Gore, to use his "great power, both legal and moral" to exercise control over the prices of an entire industry, and to initiate or participate in a whole series of administrative and legislative action of a punitive nature if that control were not accepted.

The President, according to Professor Barber, agreed ahead of time to support the position of the senators and did so in response to a prearranged press conference question on steel prices.

As indicated above, the next part of the program was to write letters to the chief executives of eleven steel companies and to the Steelworkers Union. The response made to the letter addressed to me has been referred to. I have also noted the visit to see President Kennedy at his suggestion following the exchange of letters.

The point here is to note that apparently for a number of months the President and his associates had devised a plan for government intervention into the automobile and steel negotiations. This plan, although Presidential control of wages and prices had not been authorized by Congress, had appeared to them to work reasonably well until it came to the matter of steel prices. In fact there was no "social contract" in steel and no commitments against the price increases following wage negotiations. Professor Barber seems to concur in this. Every indication short of saying that United States Steel would have to raise prices was given. All that had been overlooked in the great anxiety to achieve the semblance of a "social contract" or, in the language of Walt Rostow, a "treaty," in the steel industry, which, of course, would have involved squeezing further an already low profit steel industry.

I have noted before that to have had any kind of agreement among competitors and a single union with respect to prices would have been, in my judgment, improper. Attempted government intervention does not change that situation unless the laws are changed and the United States moves away from being a competitive society. Instinctively the Council of Economic Advisers also recognized that it was undesirable for labor and management to "bargain implicitly about the general price level."[9]

The significance of all the behind-the-scenes maneuvering by Administration forces was simply not comprehended by those negotiating the steel contract on the company side. Those maneuverings, however, may help to explain President

9. Economic Report of the President, 1962, p. 188.

Kennedy's reaction to the steel price increase and make his reaction more understandable.

IX. The Tidal Wave of Official Pressure

Certainly never before in the nation's history have so many forces of the federal government been marshaled against a single American industry as were arrayed against steel. And this because eight companies in that industry, acting individually and solely in their own interests, tried to increase their prices in the hope of bolstering sagging profits and thus make it possible to accelerate the modernization of their productive equipment and strengthen their own operations in the intensive competition for markets that is the central, every day fact of steel's business life.

On Saturday, April 14, 1962, the headlines on the first page of *The New York Times* said, "Steel Gives In, Rescinds Rises Under Pressure by the President: He Says Decision Serves Nation."

The subheadlines read: "Kennedy is Victor Uses His Full Powers for 72 Hours to Subdue Industry."

And that is what happened. A chronological account of the episode was printed in *The New York Times* of April 23, 1962.[10] Excerpts from *The New York Times* and other papers are in Exhibit F.[11]

All in all it was an horrendous three days of what the joint Senate-House Republican leadership charged was a "punitive, heavy-handed and frightening use of naked political power." Being on the receiving end, this looked like a rather apt description.

The Republican leadership statement issued on April 19, 1962 also said that the price controversy raised a fundamental issue: "Should a President of the United States use the enormous powers of the Federal Government to blackjack any segment of our free society into line with his personal judgment without regard to law."

As I look back, it was a period of intense pressure—many telephone calls; conferences of United States Steel executives;

10. Exhibit E. *See page 130.*
11. Exhibit F. *See page 142.*

reports of the government campaign to rouse support from members of Congress and from Democratic governors; newspaper reports of FBI investigations; a 3:00 A.M. FBI call on a reporter in the Philadelphia Bureau of Associated Press, compared by the chairman of the Republican National Committee with the "knock on the door" techniques of Hitler's Gestapo; the hundreds of calls made by a large host of politicians enlisted to assist the Administration; the demands made by the Administration, both in and out of the steel industry, by telephone and in person, were repeated and insistent; grand jury investigations; shifting of government business; pressure on competitors who felt obliged to follow the will of the Administration regardless of what good business economics or the national interest dictated; many calls from all forms of the media; a hurriedly arranged press conference, as well as meetings with Washington callers; preparation for stockholder contacts; to say nothing of the anxiety of members of my family annoyed by calls late at night and distressed by my public opprobrium, or the men who wore "S.O.B." buttons in response to President Kennedy's remark about businessmen. All in all it was an experience to be remembered.

It may be useful to recall some specific events touching United States Steel and other companies in the steel industry.

The Department of Justice statement on the night of April 10, 1962, that the Department would make an inquiry into the price increase, was followed by an announcement from Attorney General Robert F. Kennedy that he had ordered a federal grand jury investigation in New York and had subpoenaed price records of United States Steel. A bit later, similar records of Bethlehem, Jones & Laughlin, Armco Steel Corporation, and National Steel were subpoenaed. Also, the Department of Justice said it was studying whether United States Steel "so dominates the industry that it controls prices and should be broken up."

Two days after all the companies that had increased their prices had revoked that action, the Justice Department subpoenaed records of Youngstown Sheet & Tube Company, Inland Steel Company, Kaiser Steel Corporation, Colorado Fuel & Iron Corporation, Wheeling Steel, and McLouth Steel Corporation. Of the companies commanded by the Department to submit records, Armco, Inland, Kaiser, Colorado Fuel & Iron, and McLouth did not announce increases in their prices.

At the end of May 1962, *The New York Herald Tribune* reported that, in the first step of this kind in many years, the Justice Department had, on May 18, subpoenaed certain records of the American Iron & Steel Institute.

Representative Emanuel Celler (Democrat, New York), as chairman of the House Judiciary Subcommittee on Antitrust and Monopoly, announced on April 11, 1962 that the Subcommittee would hold early hearings on the steel price increase. Some time after the price increases were rescinded, Mr. Celler is reported to have said that his Subcommittee's investigation had been called off.

Also on April 11, Paul Rand Dixon, counsel to the Kefauver Committee when it conducted a lengthy investigation of steel prices and profits in 1957, and later chairman of the Federal Trade Commission, said that the FTC had begun an informal inquiry into the price increases that could lead to penalties of $5,000 a day for violation of the Commission's consent decree of June 15, 1951.

Senator Kefauver moved right ahead on the investigation he had promised would be undertaken by the Subcommittee on Antitrust and Monopoly of the Senate Judiciary Committee. On April 12 the Subcommittee issued subpoenas requiring twelve steel companies to produce voluminous basic cost data for the years 1954 and 1961. After numerous conferences between the Subcommittee staff and officers of United States Steel, this demand was substantially reduced.

Subsequently four companies, Republic, National, Armco, and Bethlehem, refused to comply with the modified request on the ground that the subpoenas were beyond the power of the Kefauver Subcommittee. By a vote of five to two, the Subcommittee recommended to the Judiciary Committee that these companies and a number of their officials be cited for contempt. After a series of hearings, the Judiciary Committee refused by a vote of ten to five to uphold the recommendation, and as a consequence all twelve companies were released from any obligation to provide the cost data the Subcommittee had called for.

Under Washington dateline of April 13, 1962, the Associated Press carried a dispatch saying, "The Pentagon today ordered defense contractors and their suppliers to shift steel purchases to those companies which have not raised prices." As an example of this, according to the *Herald Tribune*, the Defense Department announced on April 13 "that instead of buying some steel from U. S. Steel for Polaris submarines, it would buy all 11,000 tons of a new order from Lukens Steel Company of Coatesville, Pennsylvania. Lukens did not raise its prices." That story also reported that Secretary of Defense Robert S. McNamara had declared that "the Pentagon and other Government agencies would shift their orders to various companies that did not boost their prices."

The steel involved in the 11,000-ton order for the Polaris program was HY-80 steel, a higher strength steel developed by United States Steel specifically for submarines operating at great depths. The Associated Press said the order amounted to $5- to $6-million.

In the United States Steel news conference of April 12, 1962, I was asked to comment on the President's statement, in his news conference of the day before, that Defense Secretary McNamara had calculated that a $6 per ton rise in the price of steel would add $1 billion to the nation's defense budget. I said that if the increase were applied to both direct and indirect purchases of steel for defense purposes, the additional outlay would be "something in the nature of $20 million. And where the 'billion dollars' comes from, I don't know, unless someone made a projection that this in some way would extend to other things."

The editor of *The Shreveport* (Louisiana) *Times* did some subsequent research on this point and came up with this accurate finding:

Production of raw steel in 1961 was about 100 million net tons for the entire industry. But because of a reduction due to cropping, trimming, and other sizing operations, actual shipments of finished steel products in 1961 totalled only 66,126,000 tons—the only tonnage of steel production from which revenue is obtained.

At a $6-per-ton increase, the increased cost of *all* steel sold by the industry would have been $396,756,000—a little more than one-third of $1 billion.

The Shreveport newspaper found that the best estimates of the amount of steel actually used for defense purposes is about

ninety-seven

3 million to 3½ million tons annually. The $6 increase per ton would, therefore, have meant some $18- to $20-million each year—$980 million less than $1 billion.

I do not, of course, pretend to know all that went on in Washington after United States Steel's price rise was announced, but *The New York Times* published very detailed accounts of official activity that were brought together later (April 23, 1962) in an almost hour-by-hour recapitulation of events at the peak of the price controversy.[12]

The *Times* reported that, after the news of an upward price movement in steel had broken, "the strategy" of the Administration had been "to divide and conquer." If two major companies could be persuaded to hold the line, the rest would have to retreat.

The key target in the strategy, the *Times* report continued, was Inland Steel, eighth largest in the industry. A secondary target was Armco Steel, the sixth largest.

> Behind all the public declaiming and swinging of clubs, Administration officials were engaged in intense personal campaigns to persuade these two not to follow the lead of United States Steel, and to persuade other companies to retreat if possible.
>
> By long distance telephone the ranking powers of the Federal Government called the ranking powers of the steel industry—management, directors, and stockholders. Edward Gudeman, Jr., the Under Secretary of Commerce, was a central figure. He knew Inland Steel well as a lifetime Chicagoan, and he handled the dealings there.
>
> President Kennedy turned to the long distance telephone, too.

As to whether "political pressure," in the invidious sense of that term, was brought to bear upon individuals connected with other steel companies, I have no first-hand knowledge. I can say—adapting one of the lines of the late Will Rogers—much of what I know is what I read in the newspapers. So far as United States Steel is concerned, several of my associates had conversations in which the subject of the price change was broached with varying degrees of vigor. As to the overall pressure to rescind our price increase the record speaks for itself.

At noontime on Thursday, April 12, Arthur Goldberg, as

12. Exhibit E.

Secretary of Labor, telephoned United States Steel's Finance Committee chairman, Robert C. Tyson, and invited Mr. Tyson to meet him at the earliest possible time in Washington. Mr. Tyson explained that, because of the news conference United States Steel was holding in mid-afternoon that day, it was felt desirable that he remain in New York until that conference had ended. Mr. Tyson said, however, that if it were of sufficient urgency he would go to Washington immediately. The upshot was that he flew to the Capital early in the afternoon and met Mr. Goldberg in a Washington hotel in mid-afternoon.

Mr. Goldberg, of course, supported President Kennedy's position, and felt that United States Steel's price increase should be rescinded as rapidly and gracefully as possible. The question was opened as to whether United States Steel would consider rescinding the increase immediately, pending the formation of a commission to be appointed by the President and composed of a leading jurist and public-spirited citizens representing various segments of the economy. The commission would conduct a thorough inquiry into the need for a steel-price rise; and if the commission determined that a need existed, then presumably there would be no objection to an increase.

Mr. Tyson made it clear that United States Steel could not surrender freedom of pricing to a commission, but that if the President did appoint such a body, out of respect for the office of the Presidency we would not take an aloof attitude even though the approach suggested was not one he believed consonant with the principles on which the American economy is based.

At the instance of the Administration, Mr. Tyson, while in Washington, met on the night of April 12 with Clark Clifford, distinguished attorney who had served as counsel to President Truman from 1946 to 1950. Mr. Kennedy, as was later publicly disclosed, had asked Mr. Clifford, because of the latter's familiarity with corporate problems, to act as the President's emissary in doing whatever he could to resolve the controversy over the steel price situation that was causing such a ferment in the nation.

Mr. Clifford and Mr. Tyson discussed a number of the same matters that had been discussed by Mr. Goldberg and Mr. Tyson. Mr. Clifford also concurred fully in President Kennedy's

ninety-nine

attitude, held the price increase to be an error, and believed that it should be rescinded immediately.

Mr. Goldberg and Mr. Clifford reaffirmed their positions to us the next day, as I shall presently report.

Agents of the Federal Bureau of Investigation did appear at United States Steel's offices in New York. They came during regular office hours on Thursday, April 12, 1962.

Two agents of the FBI called at my office in New York in mid-afternoon, while I was engaged in the news conference mentioned earlier. The FBI men said they wished to talk with several of our executives. An officer of United States Steel recommended that they confer with our general counsel, John S. Tennant, who was also at the news conference. Mr. Tennant talked with the FBI representatives the following day, Friday, April 13, 1962, and arranged for the government agents to see the executives they had mentioned.

Later that day, at about 5:30 P.M., one of the FBI men called Mr. Tennant by telephone and said that he had subpoenas to serve on certain executives of United States Steel for appearance before the grand jury in New York the following Monday. While it is unusual to allow so little time between the service of a subpoena and an appearance before a grand jury, arrangements were made for two of the executives to accept service on Saturday, April 14, and they later testified before the grand jury.

On Monday, April 16, Arthur Goldberg called Mr. Tennant and suggested that it might be helpful if United States Steel's general counsel could meet with Attorney General Kennedy. Mr. Tennant made an appointment to confer with the Attorney General in the latter's office in Washington that evening.

Mr. Tennant found Attorney General Kennedy relaxed and friendly. He told Mr. Tennant that the Administration was not vindictive and that there would be no further problems relating to subpoenas and grand jury proceedings. Mr. Tennant said that United States Steel believed a cooperative attitude between government and business to be highly important, that we intended to give cooperation, and certainly felt no desire to engage in recriminations.

The Attorney General then called in Deputy Attorney General Katzenbach and Assistant Attorney General in Charge of the Antitrust Division Loevinger, and said he wanted them to hear what he had told Mr. Tennant. He then repeated the substance of the previous conversation and told Mr. Tennant if he had any further problems with the Antitrust Division, he should call either Mr. Kennedy or Mr. Katzenbach. Mr. Tennant left with the definite impression that the subpoenas and grand jury proceedings were a part of the campaign to obtain a cancellation of the price increases and that since this had taken place, the grand jury proceedings would either be discontinued or allowed to fade away.

At various times shortly after the steel price controversy had subsided, the Antitrust Division of the Department of Justice subpoenaed, in connection with the grand jury inquiry, expense accounts and telephone numbers of certain executives of United States Steel, Bethlehem, Jones & Laughlin, Wheeling Steel, Youngstown Sheet & Tube, Inland, Armco, and National.

Some months later the grand jury proceedings with respect to United States Steel were discontinued.

X. Steel Prices Lowered in the Storm

After eight companies in the steel industry had announced price increases on April 10, 11, and 12, 1962, the big question was whether those increases could remain in effect, because the rest of the companies were still selling at prices averaging $6 less than those eight, and because of the manifold government pressures that were being exerted.

The question began to resolve itself about mid-morning of Friday, April 13. A Chicago newspaper reported that one of its foreign correspondents had interviewed Joseph L. Block, chairman of Inland Steel, who was traveling in Tokyo, and that Mr. Block had declared he was opposed to a price increase at that time. Around noon, Inland announced that its board of directors had formally decided not to change the level of that company's prices. The decision was announced in a company statement:

> The company has long recognized the need for improvement in steel industry profits in relation to capital invested. It believes this condition which does exist today will ultimately have to be corrected. Nevertheless, in full recognition of the national interest and competitive factors, the company feels it is untimely to make any upward adjustments.

Kaiser Steel Corporation, on the Pacific Coast, followed very shortly afterward with an announcement that it too had decided against a price increase.

Mr. Tyson, Leslie B. Worthington, United States Steel president, and I met with Mr. Clifford and Mr. Goldberg at the Hotel Carlyle in New York on Friday, April 13, 1962—a meeting about which a good deal has appeared in public print, some of it highly speculative.

We met for lunch that day, at about 1:00 P.M. We already knew that Inland Steel, which had its entire production in the Chicago area in the heart of the largest steel-consuming area in the country, and where United States Steel also has major steel works, had officially announced its decision not to increase its price. And we knew, too, of the decision against a price change by Kaiser Steel, which competed against operations of United States Steel in the West.

I believe that the three of us from United States Steel knew the significance of these circumstances better than did either Mr. Clifford or Mr. Goldberg. Having been apprised of the decision by Inland and Kaiser, we knew, before we went to the meeting at the Hotel Carlyle, that maintenance of our new price level was not commercially feasible; we knew that our price increase would not stand. I think that it is accurate to say that Mr. Clifford's request that we roll back our increase as a means of ending the controversy then engulfing official Washington and commanding widespread attention throughout America was, by that time, rather academic.

Our discussion with Mr. Clifford and Mr. Goldberg had been progressing for perhaps the better part of two hours, when each side to the discussion was notified by telephone that Bethlehem Steel had rescinded the price increase it had announced the morning of April 11. Bethlehem's announcement meant that the company second largest to United States Steel in our industry, would immediately be taking orders at prices averaging $6-per-ton lower than ours were at the moment. This was the clincher, if any was needed.

No one has yet discovered a way to sell steel at $6 a ton more than the competition in times of ample supply. Accordingly, United States Steel prepared the following announcement and released it to press, radio and television shortly before 5:30 P.M.

on Friday, April 13, 1962:

United States Steel Corporation today announced that it had rescinded the 3½ per cent price increase made on Wednesday, April 11. (The price increase we had announced the evening of April 10 had taken effect at 12:01 A.M. the next day.)

The price recision was made in the light of the competitive developments today, and all other current circumstances including the desire to remove a serious obstacle to proper relations between government and business.

The announcement of the Bethlehem cancellation, which had preceded ours by about two hours, was contained in a company statement saying,

Although we still hold the opinion that a steel price increase is needed under present conditions to insure reasonable earnings to provide the funds necessary to build more competitive facilities— and at least partly offset the past employment cost increases which have been absorbed without a price increase—we must remain competitive.

The night of April 13, 1962, Jones & Laughlin announced that it too was cancelling its price increase, as did Republic which made this statement:

We take this action to remain competitive even though it is our sincere belief that a price increase is needed at this time to offset to some extent the increased costs we have incurred since 1958.

Price cancellations then followed in rapid order by Pittsburgh Steel, National Steel, and Youngstown Sheet & Tube, before April 13 had expired. The next day, April 14, 1962, Wheeling Steel cancelled its price rise.

XI. The Public Reaction

We in United States Steel were surprised, and those in Washington must have been also, by the vigor and volume of expressions from individual citizens, many of whom, including some who censured us for trying to increase our prices, said they were displeased with the tone and substance of the President's heated reaction to our attempt to raise prices.

In the days following our price announcement and President Kennedy's reaction, there was a great influx of letters, telegrams

one hundred three

and postal cards to officers of United States Steel. Of these, 2,200 upheld United States Steel's position or were critical of the position taken by President Kennedy. About 900 were critical of United States Steel or concurred in the President's stand. About 60 communications could not be actually classified as either pro or con. On April 17, 1962 a press dispatch from Washington reported that the reaction received by the White House was against the steel price increase by a margin of five to two.

We made an analysis of press comments on the price controversy by sampling periodicals, columnists' articles, cartoons, and letters to the editor that appeared in the nation's major newspapers and periodicals from April 11 through May 6, 1962.

A total of 1,346 comments were investigated, of which 960 expressed definite views for or against either United States Steel or the Administration. Of these 960 items, 592—or more than 60 per cent—were favorable to United States Steel.

Letters to the editor were the principal source of pro-United States Steel or anti-Administration sentiment. They ran more than three to one in our favor. Columnists' articles supported our position by a three to two margin. More than half the editorials (55 per cent) were favorable to United States Steel. Cartoons alone tended to support the Administration by a ratio of seven to five.

Reaction immediately after the announcement of our price increase on the evening of April 10, 1962, and after President Kennedy's news conference the next afternoon, tended to be generally unfavorable to United States Steel. A change in sentiment became strongly evident as early as April 14, 1962. From that point on, the trend of press and public comments became increasingly in support of the position United States Steel had taken, so that in the closing weeks of the study opinion was running three to one in that direction.

A cross section of statements by leaders in public office, politics, labor and business has been included in the exhibits.[13]

13. Exhibit G—Cross section of opinion. *See page 147.*

XII. The Media—And My Response
To President Kennedy

This recounting would not be complete without a mention of my response to President Kennedy and the activities of the media.

Recognizing the potential of the President Kennedy-steel story, the media pulled out all the stops and that was to be expected. Overnight, radio, television, newspapers and magazines moved swiftly in a manner best described to those interested in hunting as "hot pursuit."

Suddenly I became one of the most notorious men in America. Many people recognized me on the street and would stare at length as if to say "What manner of man is this who has deigned to incur so much Presidential displeasure?" The atmosphere created was almost one of belligerency. It was quite an experience to be trailed from a New York City apartment door to the office of United States Steel at 71 Broadway, downtown, at 7:00 A.M. by an ambitious reporter.

After the President spoke there was a clamor for a press conference. Busy as we were, and for a few days we were extremely engaged, one was arranged for Thursday, April 12, the day after the President's press meeting. About 150 news and camera men attended.

The text of my opening statement at that conference, together with the transcript of the question-and-answer period that followed is included in the exhibits.[14]

As I said in the *Look* magazine article of January 29, 1963, I decided on a tempered answer, although it would have relieved a lot of tension and frustration if I had yielded to natural impulses as others in industry and elsewhere did later. But that was not my way. I knew an angry answer to a heated denunciation would serve only to widen the rift between government and business which was already becoming so apparent.

In my statement I said,

> When the President of the United States speaks as he did yesterday regarding our Corporation and its cost-price problems, I am sure a response is indicated and desirable.

14. Exhibit H. *See page 150.*

Let me say respectfully that we have no wish to add to acrimony or to public misunderstanding. We do not question the sincerity of anyone who disagrees with the action we have taken. Neither do we believe that anyone can properly assume that we are less deeply concerned with the welfare, the strength and the vitality of this nation than are those who have criticized our action.

I said that the nation's strength depended on keeping its productive machinery and equipment in good order, and that to do this, our company, in common with others in industry generally, must be profitable. I sketched the role of profit in creating and securing jobs, and cited, as I have in this account, statements I had previously made on the need to remedy United States Steel's cost-price relationship. And, with respect to misconceptions, I pointed out that the President said, when questioned regarding any understanding not to increase prices, "We did not ask either side to give us any assurance, because there is a very proper limitation to the power of the government in this free economy."[15]

As to the economic effects of our price increase, I said that a rise of three-tenths of a cent per pound in the price of steel "adds almost negligibly to the cost of the steel which goes into the familiar, everyday products that we use." I presented, as examples, the amount by which the price change would increase the cost of steel for the following items:

Automobiles
    Standard size $10.64
    Intermediate size... 8.33
    Compact size 6.83
Toaster .... .03
Washing Machine (wringer type).............................. .35
Domestic Gas Range (4-burner) .... .70
Refrigerator (7.7 cubic feet) .... .65
303-size Food Can .... Five One Hundredths of 1 Cent

Many members of the press who were present, I am sure, were objective; but a scanning of the questions and answers shows what my recollection recalls, a tinge of the prosecutor attitude. I suppose a certain amount of this must be anticipated if the reporter is to fulfill his duty to the public. The question is how much is necessary and how much is desirable.

15. Exhibit A.

Soon after the incident began, as indicated, it had taken on a political context and became to a degree a contest between Democratic and Republican politicians. The latter, while not necessarily supporting this particular price increase, claimed the President had acted unfortunately and without authority. This contest was early reflected in the media.

When you look at the matter after a time span, the wonder is that the media was as impartial as it was. Viewed as a whole, the newspapers, at least by number, were roughly less in support of the President than supporting him. The whole episode served as another good example of the value to the nation of having an independent press and other media. Those portions of the media and their reportorial staffs who supported President Kennedy did so, I believe, because they felt that way, and not through fear of official reprisals if they did not. Those opposed to the Presidential action certainly took their stand without regard to Administration displeasure.

On the whole the media acted with courage in a politically and economically trying situation. I say this although at times I did feel the "hot breath of the prosecutor" and felt a lack of objectivity. But, then, I suppose I was not in the best position to judge.

XIII. Tuesday, April 17, At The White House

On Tuesday, April 17, just one week after my previous visit to the White House, I again met there with President Kennedy. That meeting was, in part, a by-product of the conference we had held in New York with Clark Clifford and Arthur Goldberg. Basically, however, the meeting was initiated through Hal Korda.

At that meeting the matter uppermost in my mind was business confidence. While there was some division in the business community about the wisdom of United States Steel's action in raising prices, there was almost complete unanimity in condemning the response tactics of President Kennedy. I am certain the President was equally concerned about the loss of confidence, and for good reason, since it was clearly a matter of increasing concern to the entire nation.

It may be that market conditions would have resulted in a

one hundred seven

drop in stock prices at the time even if the steel price incident had not occurred. But that incident seemed to generate a loss of confidence which was cumulative and in a few weeks evidenced itself by a precipitant drop in stock prices. A record single-day loss of $20.8 billion on May 28, 1962 was greater than the pinnacle two-day loss in the great stock crash of October 1929. In early July, U. S. News & World Report was to present a calculation that "more than $100 billion has been erased from the value of stocks in American companies."

On April 10, 1962, United States Steel common sold for $68 a share. Two months later it was off to $50 a share.

The result of this White House meeting was as good as could have been expected. At his news conference on Thursday, April 19, 1962, President Kennedy issued a statement that reflects the atmosphere in which the meeting of April 17 had taken place.[16]

The discussion began by the President stating he had no intention of seeking controls on wages and prices. He said he had to get into "this one" because of its effect on the "wage program." He realized, he said, it leaves us with a problem, referring to the matter of business confidence.

At the meeting, the proposal by Senator Humphrey for creation of a Presidential commission to conduct an inquiry into steel industry problems and practices came up in relation to the matter of bolstering business confidence.

I expressed the belief that having such a commission for the steel industry alone would be neither wise nor useful. I possessed some knowledge of the study conducted for the government of Great Britain shortly before by the commission headed by Lord Cohen, on "Prices, Productivity, and Incomes." The Cohen Commission's report held that cost inflation— notably labor cost inflation—was at the root of price inflation. It recommended that labor cost increases there not be allowed to exceed the long-term improvement in productivity in Great Britain's economy as a whole—and that wages in any occupation reflect conditions of demand for and supply of labor for that occupation. My view was that, if anything were to be done, it should be done along the broad lines of the Cohen Commission's inquiry, but that I had some question about the usefulness of that approach.

16. Exhibit I—Kennedy statement. See page 162.

The balance of payments problem also came up at our meeting and the President asked whether it might not be constructive to set up a group from The Business Council (of which I happened at the time to be chairman) to study this whole complex matter and recommend helpful measures.

I concurred and subsequently The Business Council set up a committee of the kind suggested by the President which I also chaired. The other members of the committee were Allan Sproul, Thomas Watson, Henry Alexander, Harold Boeschenstein, Crawford Greenewalt, and David Rockefeller. The committee work was broader than balance of payments and involved our making recommendations related to tax reduction and other actions affecting the American economy generally.

This committee worked mainly with Secretary of the Treasury Douglas Dillon and his deputy, Henry Fowler, but we had a number of meetings during the following months with the President and made our recommendations directly to him.

One other matter came up for discussion involving the need for improvement in the profits of the steel companies, and there was some review of the Administration's plans with respect to depreciation. A tax bill with provision for a tax credit for industry investment in new equipment was referred to by the President. My position was that depreciation improvements would help but were not a substitute for a proper cost-price relationship, one that would tend to generate real economic progress and build the competitive sinews American industry needed.

As an outgrowth of this, officials of United States Steel held discussions with Secretary Dillon and his associates on the depreciation question, to good effect, I believe. Undoubtedly for many reasons, as well as the steel situation, the rates of depreciation were subsequently changed by Secretary Dillon who well understood the need.

Incidentally, early in the discussion the President said "Bobby" would cooperate on the grand jury matter; that he, the President, would talk with Representative Celler regarding his investigation, but had less influence with Senator Kefauver with respect to his hearings.

Only as I was leaving did the matter of price come up. The President inquired as to our further intentions in this area. I replied that we had no immediate intention to announce increases but the need was still there as he knew. His statement, as I recall it, was that he hoped there would be no surprises.

XIV. Aftermath

A decree from high places on price matters does not cure the economics of higher and higher costs. In bits and pieces over the next twelve to eighteen months, many of the price changes originally attempted were made, publicly announced, and generally sustained by the marketplace.

In April 1963, after some steel companies had announced price increases on selected steel products, President Kennedy was prodded into making a public statement that he would act if steel price increases became general. However, in this, and in subsequent price changes, the "heat" seemed to have been dissipated and President Kennedy and his associates appeared to recognize that against mounting costs, price edicts have their limitations.

Other presidents in other times have also had involvement in steel pricing. In October 1964 President Lyndon Johnson said in substance that steel price increases were not in the national interest. Some called it "jawboning." Others thought of it as Executive usurpation of the marketplace in the Kennedy tradition.

In January, 1965, the Council of Economic Advisers was requested by President Johnson to study recent steel price increases and the Council reported that the industry did not need to raise prices. My recollection is that the industry at that point was just about last, that is, among the lowest from the standpoint of profitability, on a list of about thirty-nine industries compiled by the First National City Bank of New York. With all due respect, the Council of Economic Advisers seemed again to address itself to a political answer rather than an economic one.

In January 1966, as I recall it, Bethlehem Steel Corporation raised prices on structural steel $5 a ton—a move which President Johnson immediately objected to as an "unwarranted price increase that could lead to inflation." The Defense Department, Commerce Department, and the General Services Administration were instructed to place orders for structural steel with the companies which had not raised their prices.

On this occasion I was called to Washington and met with Secretary of Defense Robert McNamara. Before going, we reviewed the situation at length internally in United States Steel. It was our commercial judgment that the marketplace would not sustain that much of an increase and that we should follow what we thought to be a correct commercial judgment of raising products in this classification an average of about $2.75 a ton. On meeting with Secretary McNamara I explained our conclusion and the reasons for it. He seemed satisfied and shortly after that we announced the price changes we had in mind which were subsequently followed by others in the industry.

In August 1966 (following a wage settlement in which President Johnson played an active role) a price increase was announced for some flat-rolled products. This modest increase of about 2 per cent was declared by the White House to be irresponsible and inflationary, but the increase stood.

The same thing happened in 1967 when, following price announcements in steel, the chairman of the Council of Economic Advisers sent telegrams to major steel companies which had not raised prices urging them to hold the price line. President Johnson also expressed regret over the price changes. Near the end of the year the President announced that the United States would take steps to stem such increases.

In August 1968 Bethlehem Steel Corporation announced a 5 per cent general price increase to offset employment cost increases. President Johnson again reacted, instructing federal agencies to ban purchases from companies that had announced increases. After internal deliberation, United States Steel announced price increases approximately half of those previously announced by others.

In July 1969, again to offset rising costs, price increases on a wide variety of products approximating 4.8 per cent were made. This increase was attacked by members of the Senate Economic Committee and the House Banking Committee, and restoration of guidelines was requested with a request for formal controls in the background.

In January, 1971, Bethlehem Steel Corporation announced a 12 per cent price increase of structural steel items. President Nixon condemned the action and the White House indicated that it might favor increases in steel imports if other companies followed Bethlehem. Increases approximately half that size were announced by other companies and the Administration indicated their acceptability.

In August 1971, following labor negotiations, price increases were announced. These were also criticized by the White House. Shortly afterward, although undoubtedly for other reasons, the Nixon Administration imposed the Phase I wage and price freeze.

During the control period the steel industry, like many other industries, was required to function under price criteria which resulted in some companies being permitted to charge much higher prices than other companies. This period of price controls also resulted in squeezing the profit margins of many companies. Formal controls ended April 30, 1974, to be followed by catch-up price increases in many industries, including steel.

In December of 1974 United States Steel announced some price increases which again were frowned upon, this time by the Gerald Ford Administration. Dr. Albert Rees, acting as chairman of the legally constituted Council of Wage and Price Stability, requested price information which was furnished. Subsequently, after negotiations, United States Steel's price increases were "rolled back" approximately 20 per cent according to Dr. Rees' announcement.

It is pertinent to note that except for this, and the period during 1971 to 1974 of formal controls, other price control actions taken during the administrations of President Kennedy, President Johnson, and President Nixon were extralegal. Nevertheless, action was taken and in many cases quite vigorously pursued as I have noted. Other industries have also been subjected to Presidential price attention, for example, the aluminum industry in 1965.

The steel industry is a capital intensive industry. This is to say that large investments of capital are required for productive facilities and those investments must be made on a long term basis. Many investments involve plant and equipment designed to operate twenty to forty years. It can frequently take three or four years from the time of authorizing a major steel project until it is fully engineered, the machinery and equipment built, and the project finally constructed and in full operation after a start-up period.

The political decisions which activated the Executive branch against price increases here recited were usually short-run decisions, although this conclusion may be considered by some as a debatable one. I have reached this conclusion, however, after watching the process a number of years and commend it for careful scrutiny.

When in 1973 and in 1974 shortages of domestic steel production became quite evident, the question can properly be asked whether those shortages were the responsibility of public officials rather than the responsibility of the individual companies in the steel industry.[17]

It is understandable that in an election year, denouncing a steel price increase seems to be good politics. What happens in that near election is one thing. What happens to any industry subjected to this type of treatment in a competitive world is quite another.[18]

XV. Reappraisal

Participants in major public issues, myself included, are sometimes unable to see all the implications of the events in which they have been enveloped. Here I have, in unhurried retrospect and reassessment, tried as best I can to give a reasonable account of the steel price ordeal of mid-April 1962.

These few observations may be added.

Pricing by political pressure, rather than pricing by the forces exerted in competitive markets, has serious economic and practical limitations. The marketplace is not a kindly arbiter as any corporate official knows, but it is much less likely to err than the official hop, skip, and jump decision making of those unskilled in the nuances of their own decisions.

Moreover, it is difficult indeed, and understandably so, for those designated or selected for political functioning to reach a non-political conclusion. And politically entangled conclusions are simply not satisfactory to accomplish desirable economic ends upon which the welfare of the State depends.

17. See "The Anatomy of a Steel Shortage" (10/29/74) by M. Tenenbaum, President, Inland Steel Company.
18. For a general discussion of government and business relations see "The Washington Embrace of Business" by Roger M. Blough, a series of three lectures at Carnegie-Mellon University, November 1974.

The World-Telegram of May 7, 1962, referring to the United States Steel stockholders' meeting held that day, said in part:

Mr. Blough labeled 'as incomprehensible' the concept that a 3½ per cent steel price rise 'somehow constitutes an intolerable threat to the security and economic welfare of this nation.' After all, the chairman continued, in the four years since the last steel price hike, the price of many other important industrial products has gone up and so has the federal budget—'or the cost of government if you will—24 per cent in the last four years with 9.3 per cent of that increase in the past year.'

Although the Council of Economic Advisers counseled President Kennedy that the steel industry would have "good profits" without price increases, the cost squeeze continued and deepened after the price episode. As a result, many of the very modest dividends then prevailing in the steel industry were cut.

In July 1962 Lukens Steel cut its dividend from 40 cents a share to 25 cents. In August, Republic Steel cut from 75 cents to 50 cents, and Wheeling Steel cut its dividend in half, from 50 cents to 25 cents.

Of the nine largest steel producers, four failed to earn the dividends they paid for the second quarter of 1962 and three others earned their dividends by too small a margin.

The following data regarding the shipments and earnings of the nine largest steel companies is helpful to understand the cost squeeze predicament of the steel industry in 1962.

Nine Largest Companies

	1957	1958	1961	1962
Shipments	64.5M	48.25M	51.2M	53.3M
Sales	12.2B	9.8B	10.2B	10.7B
Net Income	964.2M	692.5M	590.1M	473.6M
Per Cent Return on Sales	7.9%	7.0%	5.8%	4.4%

As the data shows, shipments moved up in 1962 by 10 per cent from 1961, while return on sales moved down 37 per cent. Incidentally, employment costs moved up from 1958 to 1962 by 18.3 per cent.

Some may ask, "What difference does it make?" One of the smaller companies in steel production in 1962 was Pittsburgh Steel Company, and it had an answer to that question. On May

24, 1962, its president, Allison Maxwell, in a talk entitled "Steel's Profit Problem" said:

> ... While prices must be low enough to hold today's market, they must be high enough to build the markets of tomorrow. Prices have a double role to play in competition. They must be attuned to the immediate; and they must also help provide profits to buy superior tools for lower costs and competitive prices in the long range. Prices reflect a delicate balance of both short and long range competitive requirements, far too intricate for manipulation by the heavy hand of government.
>
> The cost-price squeeze on profits has become increasingly intense over the past 12 years. Undeniably, we have been gaining productivity. If we didn't we would be presiding at our liquidation. But the gains have been buried by wage increases that outstrip productivity advance by more than 9 to 1.
>
> Even the latest agreement exceeds by 50 per cent the average annual productivity increase since 1940. And this agreement does not become 'non-inflationary' just because the Administration puts that label on it. It is not a non-inflationary agreement, just because it is less inflationary than previous agreements. It is, in fact, just one more increment in the inflationary trend that has been long developing. . . .
>
> More productivity gains must carry through to profit, so that productivity will continue to gain—and gain by leaps and bounds—if we are to surpass our competition. All answers to the threat by competition revolve around rapid improvement of production tools and unrestricted freedom to market our products profitably.
>
> For some unexplainable reason, these two inseparable concepts do not enjoy equal popularity. Nearly everyone will agree to the need for new and better equipment. Proposals to spur capital expenditures win popular acclaim. But suggest that industry must generate more profit to build new plants, and this arouses controversy.

Critics may say: Yes, but you do not understand the "big picture," meaning the national interest as perceived by those in authority at the moment. Perhaps some of us do not have all the facts those in officialdom have on a given subject. Nevertheless, I would ask them to remember that the "big picture" is not do-able without an industrial and service base. And the "big picture" industrially is composed of many, many smaller "pictures." It is within the employment capability and financial health of these smaller "pictures" where lies the answer to the total strength of America. Pricing by ukase is probably the "wrongest" answer one can rely upon to achieve that end. In the April 16, 1962 edition of the *Christian Science Monitor*, Erwin D. Canham, editor, asked these questions:

> After this display of naked power, whatever its provocation or justification, how free will the American economy be? And second, how will the American economy accumulate the capital to modernize and expand its plant and equipment, thus putting more people to work and competing more effectively with other countries?

When asked, in her terminal illness, "What is the answer?," Gertrude Stein is quoted as replying, "What is the question?" Thirteen years later, these questions of Erwin Canham are still good questions.

An even larger question involves the nature of the Office of the Presidency, its relations with Congress, with business, labor unions, and other segments of society, and the future of the "rule of law" versus the exercise of Executive power in this nation. It is probably one of our most complicated "gray area" unfinished pieces of business, but one which will need all the attention we can give it—before too long.

APPENDIX

EXHIBIT A *The New York Times* April 12, 1962, page 20

1

Q.—Mr. President, the unusually strong language which you used in discussing the steel situation would indicate that you might be considering some pretty strong action. Are you thinking in terms of requesting or reviving the need for wage-price controls?

A.—I think that my statement states what the situation is today. This is a free country.

In all the conversations which were held by members of this Administration and myself with the leaders of the steel union and the companies, it was always very obvious that they could proceed with freedom to do what they thought was best within the limitations of law.

But I did very clearly emphasize on every occasion that my only interest was in trying to secure an agreement which would not provide an increase in prices, because I thought that price stability in steel would have the most far-reaching consequences for industrial and economic stability and for our position abroad, and price instability would have the most far-reaching consequences in making our lot much more difficult.

When the agreement was signed—and the agreement was a moderate one and within the range of productivity increases—as I've said, actually, they'll be reduction in cost per unit during the year—I thought, I was hopeful, we'd achieved our goal.

Now the actions that will be taken will be—are being now considered by the Administration. The Department of Justice is particularly anxious, in view of the very speedy action in other companies, who have entirely different economic problems facing them than did United States Steel—the speed with which they moved the . . . it seems to me to require examination of our present laws and whether they're being obeyed by the Federal Trade Commission and the Department of Justice.

And I am very interested in the respective investigations that will be conducted in the House and Senate, and whether we shall need additional legislation, which I would come to very reluctantly.

But I must say the last twenty-four hours indicates that those with great power are not always concerned about the national interest.

2

Q.—Mr. President?

A.—Yes.

Q.—In your conversation with Mr. [Roger M.] Blough [chairman of the board of United States Steel] yesterday, did you make a direct request that this price increase be either deferred or rescinded?

A.—I was informed about the price increase after the announcement had gone out to the papers. I told Mr. Blough of my very keen disappointment and what I thought would be the most unfortunate effects of it.

And, of course, we were hopeful that other companies who, as I've said, have a different situation in regard to profits and all the rest than U. S. Steel—they're all—have a somewhat different economic situation—I was hopeful, particularly in view of the statement I saw in the paper by the president of Bethlehem, in which he stated—though now he says he's misquoted—that there should be no price increase—and we are investigating that statement —I was hopeful that the

one hundred seventeen

others would not follow the example, that therefore the pressures of the competitive marketplace would bring United States Steel back to their original prices—but the parade began.

But it came to me after the decision was made. There was no prior consultation or information given to the Administration.

5

Q.—Mr. President, if I could get back to steel for a minute. You mentioned an investigation into the suddenness of the decision to increase prices. Did you—is the position of the Administration that it believed it had the assurance of the steel industry at the time of the recent labor agreement that it would not increase prices?

A. We did not ask either side to give us any assurance, because there is a very proper limitation to the power of the Government in this free economy.

All we did in our meetings was to emphasize how important it was that the—there be price stability, and we stressed that our whole purpose in attempting to persuade the union to begin to bargain early and to make an agreement which would not affect prices, of course, was for the purpose of maintaining price stability.

That was the thread that ran through every discussion which I had, or Secretary Goldberg.

We never at any time asked for a commitment in regard to the terms—precise terms—of the agreement from either Mr. [David J.] McDonald [president of the United Steelworkers of America] or from Mr. Blough, representing the steel company, because, in our opinion, that —would be passing over the line of propriety.

But I don't think that there was any question that our great interest in attempting to secure the kind of settlement that was finally secured was to maintain price stability, which we regard as very essential at this particular time.

That agreement provided for price stability up to yesterday.

7

Q.—Mr. President?

A.—Yes.

Q.—In your statement on the steel industry, sir, you mention a number of instances which would indicate that the cost of living will go up for many people if this price increase were to remain effective. In your opinion does that give the steel workers the right to try to obtain some kind of a price—or a wage increase to catch up?

A.—No. Rather interestingly, the last contract was signed on Saturday with Great Lakes, so that the steel union is bound for a year. And of course I'm sure would have felt—going much further if the matter had worked out as we had all hoped. But they've made their agreement and I'm sure they're going to stick with it. It does not provide for the sort of action you suggest Mr. [Edward P.] Morgan [American Broadcasting Company]?

8

Q.—Still on steel, Senator [Albert] Gore [Democrat of Tennessee] advocated today legislation to regulate steel prices somewhat in the manner that public utility prices are regulated, and his argument seemed to be that the steel industry had sacrificed some of the privileges of the free market because it wasn't really setting its prices on a—on a supply and demand but what he called administered prices.

Your statement earlier and your remarks since indicate a general agreement with that kind of approach. Is that correct?

A.—No, Mr. Morgan. I don't think that I'd stated that.

I'd have to look at—and see what Senator Gore had suggested and I'm not familiar with it.

What I said was we should examine what can be done to try to minimize the impact on the public interest of these decisions, but though we had of course always hoped that those involved would recognize that. I would say that what must disturb Senator Gore and Congressman [Emanuel] Celler [Democrat of Brooklyn] and others—Senator [Estes] Kefauver [Democrat of Tennessee]—will be the suddenness by which every company in the last few hours, one by one, as the morning went by, came in with their almost, if not identical, almost identical price increases, which isn't really the way we expect the competitive private enterprise system to always work.

10

Q.—In connection with the steel situation again, is there not an action that could be taken by the Executive Branch in connection with direct procurement of steel under the administration of the Agency for International Aid—I mean the aid agency? For example, I think the Government buys about 1,000,000 tons of steel. Now, could not the Government decide that only steel—steel should be purchased only at the price, say of yesterday rather than today?

A.—That matter was considered, as a matter of fact, in a conversation between the Secretary of Defense and myself last evening. At that time we were not aware that nearly the entire industry was about to come in, and therefore the amount of choice we have is somewhat limited.

Q.—Too, on this thing, in the case of identical bids, which the Government is sometimes confronted with, they decide to choose the smaller business unit rather than the larger.

A.—I'm hopeful that there will be those who will not participate in this parade and will meet the principle of the private enterprise competitive system in which everyone tries to sell at the lowest price commensurate with the—their interests. And I'm hopeful that there'll be some who will decide that they shouldn't go in the wake of U. S. Steel. But we'll have to wait and see on that, because they're coming in very fast.

11

Q.—Mr. President, two years ago, after the settlement, I believe steel prices were not raised. Do you think there was an element of political discrimination in the behavior of the industry this year?

A.—I would not—and if there was, it doesn't really—if it was—if that was the purpose, that is comparatively unimportant to the damage that—the country's the one that suffers. I—

Q.—Mr. President?

A.—If they do it in order to spite me it really isn't so important.

Q.—Mr. President?

A.—Yes.

one hundred nineteen

12

Q.—To carry a previous question just one step further, as a result of the emphasis that you placed on holding the price line, did any word or impression come to you from the negotiations that there would be no price increase under the type of agreement that was signed?

A.—I will say that in our conversations that we asked for no commitments in regard to the details of the agreement or in regard to any policies with the union or the company.

Our central thrust was that price stability was necessary and that the way to do it was to have a responsible agreement, which we got.

Now, at no time did anyone suggest that if such an agreement was gained that it would be still necessary to put up prices.

That word did not come until last night.

14

Q.—Mr. President, the steel industry is one of a half-dozen which has been expecting a tax benefit this summer through revision of the depreciation schedules. Does this price hike affect the Administration's attitude?

A. Secretary [Douglas] Dillon and I discussed this morning, of course, all this. The matter is being very carefully looked into now.

THE WHITE HOUSE
WASHINGTON

September 6, 1961

Dear Mr. Blough:

I am taking this means of communicating to you, and to the chief executive officers of 11 other steel companies, my concern for stability of steel prices.

In the years preceding 1958, sharply rising steel prices and steel wages provided much of the impetus to a damaging inflation in the American economy. From the beginning of 1947 to the end of 1958, while industrial prices as a whole were rising 39 percent, steel mill product prices rose 120 percent. Steel wage rates also rose rapidly, causing employment costs per ton of steel to rise by about 85 percent. The international competitive position of American producers was impaired, and our balance of payments was weakened. Our iron and steel export prices from 1953 to 1958 rose 20 percent more than the export prices of our principal foreign competitors, and our share of world exports of iron and steel fell from 19 percent to 14 percent.

Since 1958, our price performance has substantially improved. Steel prices have been stable since 1958, as has the Wholesale Price Index. Industrial prices have not risen since 1959. The rise in consumer prices has been held within tolerable limits.

This record of price stability was purchased, however, at the cost of persistent unemployment and underutilized productive capacity. In the steel industry itself, the rate of utilization of capacity for the last three years has averaged under 65 percent. In consequence of our recent price experience, many persons have come to the conclusion that the United States can achieve price stability only by maintaining a substantial margin of unemployment and excess capacity and by accepting a slow rate of economic growth. This is a counsel of despair which we cannot accept.

For the last three years, we have not had to face the test of price behavior in a high-employment economy. This is the test which now lies ahead.

Under the collective bargaining contract signed in January 1960, steel industry wages and other employment costs will increase at the end of this month. The amount of the increase in employment costs per man-hour is difficult to measure in advance with precision. But it appears almost certain to be outweighed by the advance in productivity resulting from a combination of two factors—the steady long-term growth of output per man-hour, and the increasing rate of operations foreseen for the steel industry in the months ahead.

The Council of Economic Advisers has supplied me with estimates of steel industry profits after October 1, calculated on an assumption that prices are not increased. These estimates indicate that the steel industry will be earning 7 to 9 percent on net worth after taxes if the rate of operations is around 70 percent; 10 to 12 percent if the operating rate is at 80 percent; and 13 to 15 percent if the operating rate is at 90 percent. The steel industry, in short, can look forward to good profits without an increase in prices.

The owners of the iron and steel companies have fared well in recent years. Since 1947, iron and steel common stock prices have risen 393 percent; this is a

one hundred twenty-one

much better performance than common stock prices in general. Likewise, dividends on iron and steel securities have risen from $235 million in 1947 to $648 million in the recession year of 1960, an increase of 176 percent.

A steel price increase in the months ahead could shatter the price stability which the country has now enjoyed for some time. In a letter to me on the impact of steel prices on defense costs, Secretary of Defense McNamara states: "A steel price increase of the order of $4 to $5 a ton, once its effects fanned out through the economy, would probably raise military procurement costs by $500 million per year or more."

Steel is a bellwether, as well as a major element in industrial costs. A rise in steel prices would force price increases in many industries and invite price increases in others. The consequences of such a development might be so grave—particularly on our balance of payments position—as to require the adoption of restrictive monetary and fiscal measures which would retard recovery, hold unemployment at intolerable levels, and hamper growth. The depressing effect of such measures on the steel industry's rate of operations might in the long run more than offset the profit-raising effect of a price increase.

In emphasizing the vital importance of steel prices to the strength of our economy, I do not wish to minimize the urgency of preventing inflationary movements in steel wages. I recognize, too, that the steel industry, by absorbing increases in employment costs since 1958, has demonstrated a will to halt the price-wage spiral in steel. If the industry were now to forego a price increase, it would enter collective bargaining negotiations next spring with a record of three and a half years of price stability. It would clearly then be the turn of the labor representatives to limit wage demands to a level consistent with continued price stability. The moral position of the steel industry next spring—and its claim to the support of public opinion—will be strengthened by the exercise of price restraint now.

I have written you at length because I believe that price stability in steel is essential if we are to maintain the economic vitality necessary to face confidently the trials and crises of our perilous world. Our economy has flourished in freedom; let us now demonstrate again that the responsible exercise of economic freedom serves the national welfare.

I am sure that the owners and managers of our nation's major steel companies share my conviction that the clear call of national interest must be heeded.

Sincerely,

s / *John F. Kennedy*

Mr. Roger M. Blough
Chairman of the Board
United States Steel Corporation
71 Broadway
New York 6, New York

EXHIBIT C *Letter to the President of the United States*
September 13, 1961

UNITED STATES STEEL CORPORATION

The President of the United States   September 13, 1961
The White House
Washington, D. C.

Dear Mr. President:

May I respectfully acknowledge receipt of your letter of September 6 in which you express concern regarding inflation and steel wage and price movements.

I am certain, Mr. President, that your concern regarding inflation is shared by every thinking American who has experienced its serious effects during the past twenty years. And although we in United States Steel cannot forecast the future trend of prices in any segment of the steel industry and have no definite conclusions regarding our own course during the foreseeable future, nevertheless it should be useful to put some of the information referred to in your letter in proper cause-and-effect perspective. Moreover, your letter does raise questions of such serious import, including the future of freedom in marketing, that I feel impelled to include a word on that score also for whatever value it may be.

First let me assure you that if you seek the causes of inflation in the United States, past, present or future, you will not find them in the levels of steel prices or steel profits.

The facts, as developed by the American Iron and Steel Institute, are that from 1940 through 1960 steel prices rose 174 per cent, but the industry's hourly employment costs rose 322 per cent, or nearly twice as much. I use 1940 as a starting point rather than 1947 because during the war-affected years of 1940 through 1944 steel wages rose substantially as did the level of wholesale prices; but steel prices increased not at all. Any comparison of these trends which starts with postwar 1947 as a base, therefore obscures, rather than reveals the realities which the steel companies have had to face throughout this entire period of inflation.

In dollars and cents, wage-earner employment costs per hour worked increased from 90½ cents in 1940 to $3.82 in 1960 and far exceeded any productivity gains that could be achieved even though some fifteen billion dollars was invested in new and more efficient plant and equipment during this period. Shipments of steel per manhour worked (a measurement which overstates the gain in true productivity) improved by less than 40 per cent, in contrast to the 322 per cent rise in employment cost. Prices at higher competitive levels were the inevitable result.

Your letter speaks of the uncertainty in the amount of the employment cost increases which will occur at the end of this month, and which have sometimes been inaccurately reported elsewhere as being about 8 cents. While these costs will obviously vary somewhat among the several companies in the steel industry, the fact is that they will increase by about 13 cents per hour for wage earners; and beyond this the companies must face the added cost of adjusting the pay of many thousands of their salaried employees. In the aggregate, therefore, industry employment costs alone will rise by more than 200 million dollars a year on October 1.

So far as profits are concerned, your advisers have chosen to measure them in

one hundred twenty-three

terms of the return on reported net worth; and again I am afraid that this does more to confuse than to clarify the issue in the light of the eroding effects of inflation on investments in steelmaking facilities over the past twenty years. If we compare the fifty-cent profit dollars of today to the 100-cent dollars that were invested in our business twenty years ago, the resulting profit ratio can hardly be said to have any validity.

To be meaningful, therefore, any such comparison must naturally be adjusted for the effects of inflation, and when this is done it will be found that in the case of United States Steel (the adjusted data for the industry are not available) the return on investment over the past ten years has averaged about 3 per cent and at its highest point was 5.1 per cent.

The most useful measurement of the profit trend in a single industry, over an inflationary period, is of course profit as a percentage of sales. On this basis, it will be noted from the attached charts that profits in the steel industry have only once in the past 20 years equalled the 8 per cent level at which they stood in 1940, and have averaged only 6½ per cent in the past five years, thus demonstrating clearly that steel price increases during this period have not fully covered the rapid rise in total steelmaking costs.

In this connection it is interesting to note that following the steel strike which ended last year, an official government report was prepared for the Department of Labor by Professor Livernash of Harvard University. In this report, which analyzes the cost-price relationship in steel at length, Dr. Livernash concludes:

> Obviously while price policy can be debated in the short run, in the long run all cost increases must be met. Steel has done no more than this.

Another point of caution which should be noted in any discussion of profits on an industry-wide basis is that the industry's profit rate is merely an average— and averages can be dangerously misleading. Some companies will earn more than the average, while some may be suffering losses which they cannot sustain indefinitely. So it was in 1960 that among the thirty largest steel companies the profit rate as a percentage of sales ranged from a plus 9.3 per cent to a loss of 5.2 per cent.

Whatever figure your advisers may elect to use, however, the simple fact is that the profit left in any company, after it pays all costs, is all that there is out of which to make up for the serious inadequacy in depreciation, to repay borrowings, to pay dividends, and to provide for added equipment. If the profit is not good enough to do these things, they cannot and will not be done; and that would not be in the national interest.

So reviewing the whole picture, I cannot quite see how steel profits could be responsible for inflation especially when their portion of the sales dollar over the last twenty years has never exceeded 8 cents and is lower than that today.

As for the admittedly hazardous task which your economic advisers have undertaken in forecasting steel industry profits at varying rates of operation, let me respectfully underline the word hazardous. Moreover it might reasonably appear to some—as, frankly, it does to me—that they seem to be assuming the role of informal price-setters for steel—psychological or otherwise. But if for steel, what then for automobiles, or rubber, or machinery or electrical products, or food, or paper, or chemicals—or a thousand other products? Do we thus head into unworkable, stifling peace-time control of prices? Do we do this when the causes of inflation—in a highly competitive economy with ample industrial capacity such as ours—are clearly associated with the fiscal, monetary, labor and other policies of government?

As you note in your letter, Mr. President, the level of steel prices has not been increased in over three years—since 1958. In fact that level is slightly lower than it was two years ago, and in recent months, there have been a number of price changes, mostly reductions. There are many competitive factors which are constantly changing and which necessarily affect prices in the steel industry or any other industry. Aside from the inescapable employment and other costs, these competitive factors include competition among domestic steel producers, competition from other materials such as aluminum, glass, cement, paper and plastics, competition from foreign steel producers, and total customer demand as well as changing customer requirements, to mention only a few. The pressures of the marketplace are inexorable and cannot be disregarded by a steel company or any other company or, for that matter, cannot be disregarded by any nation which wishes to maintain its position in a competitive world.

In that connection, your letter mentions foreign competition in steel and that, of course, is of serious concern to producers in our industry. Between 1952 and 1959, for a composite of eight nations which are our principal competitors in steel, the average hourly employment cost for wage earners rose about 35 cents, compared with a rise of $1.48 in the United States. Overcoming this growing cost disadvantage is far from simple. Moreover these employment costs at the end of 1959 averaged less than $1.00 per hour in those same nations compared with $3.80 for us. These facts go a long way toward explaining why we are less competitive abroad than we were.

I am glad your letter, along with emphasizing steel prices, says:

> I do not wish to minimize the urgency of preventing inflationary movements in steel wages. I recognize, too, that the steel industry, by absorbing increases in employment costs since 1958, has demonstrated a will to halt the price-wage spiral in steel.

That "will" to do what we could to halt the wage-price spiral in steel involved thousands of employees and many companies in the serious human and financial cost of a 116-day strike in 1959. The subsequent settlement resulted in a 3½ to 3¾ per cent annual increase in our employment costs, and while this represented a substantial improvement from the corresponding 8 per cent average annual increase that had occurred throughout the period since 1940, it was still nearly twice as great as the average long-range improvement in steel shipments per manhour which amounted to only 2 per cent. So I am afraid there is little justification for the belief that improving productivity will offset the cost of the October 1 wage increase during any reasonable period of time.

With all due deference, moreover, I must note that during the past few months there have been a number of wage settlements which by no stretch of the imagination can be considered to be non-inflationary; and in spite of your own commendable concern regarding the inflationary effect of wage movements, there seems to have been—in this same period—a noticeable absence of any observable restraint by unions in their demands.

May I also respectfully observe that it is relatively easy to say that steel is a price bellwether and to draw conclusions from that belief. Steel, of course, is a basic and important commodity but where inflation is concerned, the price of steel is a symptom of the problem and not a bellwether cause. Simon Whitney, the Director of the Bureau of Economics, Federal Trade Commission, said in 1959—

> Over the long run, any psychological import attached to steel will hardly give it much more weight in the gross national product than the dollar figures show. I believe there is as much or more truth in viewing steel wages as the pace-setter for other industries.

The historical evidence would seem to disprove this bellwether theory moreover. For example, there was no increase whatever in steel prices during the years 1940 through 1944; but this did not prevent a substantial inflation in wholesale prices generally during the same period. Conversely from 1951 to

1956 there was virtually no net change in the wholesale price level, despite the fact that steel prices advanced by about 30 per cent.

Again there is little support for the theory that a steel price rise consistently pyramids. Dr. Whitney has pointed out that between 1953 and 1958...

The average steel-using item advanced only a tiny fraction of the amount that the rest of the [B.L.S. Consumer Price] index advanced.

Reducing prices or avoiding price increases has great popular appeal. As consumers we all want that. But we cannot have it both ways. We cannot have inflationary wage increases, higher taxes and other rapidly increasing costs on the one hand, and enjoy reduced and unrealistic price levels on the other, without endangering national growth and jeopardizing jobs. I of course know that you are fully aware of this, but the sooner all of us in the great American public understand that simple fact, the better.

Mr. President, I can speak for no one except my associates in United States Steel and myself on this most important subject. We share your conviction that the national interest must be heeded. We are keenly aware of inflation's danger to our institutions and to our national growth. The nation has already felt too greatly the damaging effects of inflation.

We are also equally aware of the national interest in maintaining strong, healthy industrial units operating under open market circumstances which permit freely-entered into buyer-seller arrangements. These units are the sources of jobs and the antidote for unemployment. They are the sources of research, of new products, of more things and more useful things for everyone. These industrial units and the compensation they pay to employees are the main source of our government's tax revenues. They are vital measures of our nation's strength.

Each industrial unit and every individual in it does bear, as your letter infers, a serious responsibility in maintaining the economic freedom of the country. I assure you that so far as our own corporation is concerned, we have in the past made every effort to conduct our affairs in the light of that responsibility and our purpose is to do so in the future.

Respectfully yours,
s / *Roger M. Blough*

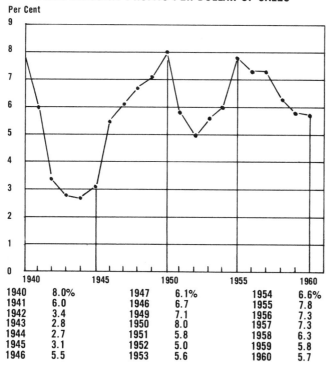

STEEL INDUSTRY PROFITS PER DOLLAR OF SALES

Per Cent

Year	%	Year	%	Year	%
1940	8.0%	1947	6.1%	1954	6.6%
1941	6.0	1946	6.7	1955	7.8
1942	3.4	1949	7.1	1956	7.3
1943	2.8	1950	8.0	1957	7.3
1944	2.7	1951	5.8	1958	6.3
1945	3.1	1952	5.0	1959	5.8
1946	5.5	1953	5.6	1960	5.7

EMPLOYMENT COSTS, PRICES AND NET INCOME
STEEL INDUSTRY

Year	Employment Costs* Per Hour Worked[1] Index 1940 = 100	Net Income as Per Cent of Total Revenue[1] Per Cent	Finished Steel Mill Product Prices[2] Index 1940 = 100
1940	100.0	8.0%	100.0
1947	172.7	6.1	130.8
1948	185.5	6.7	148.8
1949	193.7	7.1	161.1
1950	210.8	8.0	169.2
1951	233.6	5.8	182.8
1952	255.8	5.0	186.8
1953	269.6	5.6	201.0
1954	277.6	6.0	209.7
1955	300.8	7.8	219.5
1956	326.4	7.3	238.0
1957	355.4	7.3	260.6
1958	388.2	6.3	269.8
1959	419.7	5.8	274.3
1960	422.1	5.7	273.9

*Wage employees engaged in production and sale of iron and steel products
Sources: (1) Annual Statistical Reports, American Iron and Steel Institute
(2) U.S. Bureau of Labor Statistics

EXHIBIT D   *Press Release—United States Steel Corporation*
April 11, 1962

UNITED STATES STEEL CORPORATION
PUBLIC RELATIONS DEPARTMENT
71 BROADWAY, NEW YORK 6, N.Y.

FOR A.M. PAPERS
WEDNESDAY, APRIL 11, 1962

Pittsburgh, Pa., April 10—For the first time in nearly four years, United States Steel today announced an increase in the general level of its steel prices. This "catch-up" adjustment, effective at 12:01 a.m. tomorrow, will raise the price of the company's steel products by an average of about 3.5 per cent—or three-tenths of a cent per pound.

In announcing the new prices Leslie B. Worthington, President, said:

Since our last over-all adjustment in the summer of 1958, the level of steel prices has not been increased but, if anything, has declined somewhat. This situation, in the face of steadily mounting production costs which have included four increases in steelworker wages and benefits prior to the end of last year, has been due to the competitive pressures from domestic producers and from imports of foreign-made steel as well as from other materials which are used as substitutes for steel.

The severity of these competitive pressures has not diminished; and to their influence may be attributed the fact that the partial catch-up adjustment announced today is substantially less than the cost increases which have already occurred since 1958, without taking into consideration the additional costs which will result from the new labor agreements which become effective next July 1.

Nevertheless, taking into account all the competitive factors affecting the market for steel, we have reluctantly concluded that a modest price adjustment can no longer be avoided in the light of the production cost increases that have made it necessary.

If the products of United States Steel are to compete successfully in the marketplace, then the plants and facilities which make those products must be as modern and efficient as the low-cost mills which abound abroad and as the plants which turn out competing products here at home. Only by generating the funds necessary to keep these facilities fully competitive can our company continue to provide its customers with a dependable source of steel, and to provide its employees with dependable jobs. But the profits of the company—squeezed as they have been between rising costs and declining prices—are inadequate today to perform this vital function.

Our Annual Report, published last month, shows clearly the effect of this squeeze. In the three years since the end of 1958, United States Steel has spent $1 billion 185 millions for modernization and replacement of facilities and for the development of new sources of raw materials. Internally, there were only two sources from which this money could come: depreciation and reinvested profit. Depreciation in these years amounted to $610 millions; and reinvested profit, $187 millions—or, together, only about two-thirds of the total sum required. So after using all the income available from operations, we had to make up the difference of $388 millions out of borrowings from the public. In fact, during the period 1958-1961, we have actually borrowed a total of $800 millions to provide for present and future needs. And this must be repaid out of profits that have not yet been earned, and will not be earned for some years to come.

During these three years, moreover, U. S. Steel's profits have dropped to the lowest levels since 1952; while reinvested profit—which is all the profit there is to be plowed back into the business after payment of dividends—has declined from $115 millions in 1958 to less than $3 millions last year. Yet the dividend rate has not been increased in more than five years, although there have been seven general increases in employment costs during this interval.

This squeeze which has thus dried up a major source of the funds necessary to improve the competitive efficiency of our plants and facilities, has resulted inevitably from the continual rise in costs over a period of almost four years, with no offsetting improvement in prices.

Since the last general price adjustment in 1958, there have been a number of increases in the cost of products and services purchased by the Corporation, in state and local taxes, and in other expenses including interest on the money we have had to borrow— an item which has jumped from $11½ millions in 1958 to nearly $30 millions in 1961.

And from 1958 through 1961, there have been industry-wide increases in steelworker wages and benefits on four occasions amounting to about 40 cents an hour, and also increases in employment costs for other employees. These persistent increases have added several hundred million dollars to the employment costs of U. S. Steel, without regard to future costs resulting from the new labor agreement just negotiated.

In all, we have experienced a net increase of about 6 per cent in our costs over this period despite cost reductions which have been effected through the use of new, more efficient facilities, improved techniques and better raw materials. Compared with this net increase of 6 per cent, the price increase of 3½ per cent announced today clearly falls considerably short of the amount needed to restore even the cost-price relationship in the low production year of 1958.

In reaching this conclusion, we have given full consideration, of course, to the fact that any price increase which comes, as this does, at a time when foreign-made steels are already underselling ours in a number of product lines, will add—temporarily, at least—to the competitive difficulties which we are now experiencing. But the present price level cannot be maintained any longer when our problems are viewed in long-range perspective. For the long pull a strong, profitable company is the only insurance that formidable competition can be met and that the necessary lower costs to meet that competition will be assured.

Only through profits can a company improve its competitive potential through better equipment and through expanded research. On this latter phase we are constantly developing lighter, stronger steels which—ton for ton—will do more work and go much farther than the steels that were previously available on the market. They thus give the customer considerably more value per dollar of cost. As more and more of these new steels come from our laboratories, therefore, our ability to compete should steadily improve. But the development of new steels can only be supported by profits or the hope of profits.

The financial resources supporting continuous research and resultant new products as well as those supporting new equipment, are therefore vital in this competitive situation—vital not alone to the company and its employees, but to our international balance of payments, the value of our dollar, and to the strength and security of the nation as well.

EXHIBIT E *The New York Times* April 23, 1962, page 1

© *New York Times*—Used with Permission.

STEEL: A 72-HOUR DRAMA WITH AN ALL-STAR CAST

By WALLACE CARROLL

Special to The New York Times.

WASHINGTON, April 22—It was peaceful at the White House on the afternoon of Tuesday, April 10—so peaceful that the President of the United States thought he might have time for a nap or a little relaxed reading.

Just to be sure, he called his personal secretary, Mrs. Evelyn Lincoln, and asked what the rest of the day would bring.

"You have Mr. Blough at a quarter to six," said Mrs. Lincoln.

"Mr. Blough?" exclaimed the President.

"Yes," said Mrs. Lincoln.

There must be a mistake, thought the President. The steel negotiations had been wound up the previous week.

"Get me Kenny O'Donnell," he said.

But there had been no mistake—at least not on the part of Kenneth P. O'Donnell, the President's appointment secretary.

Whether Mr. Blough—Roger M. Blough, chairman of the board of United States Steel Corporation—had made a mistake was a different question.

For when he walked into the President's office two hours later with the news that his company had raised the price of steel, he set off seventy-two hours of activity such as he and his colleagues could not have expected.

Period of Excitement

During those seventy-two hours, four antitrust investigations of the steel industry were conceived, a bill to roll back the price increases was seriously considered, legislation to impose price and wage controls on the steel industry was discussed, agents of the Federal Bureau of Investigation questioned newspaper men by the dawn's early light, and the Defense Department—biggest buyer in the nation—began to divert purchases away from United States Steel.

Also in those seventy-two hours—and this was far more significant—the Administration maintained its right to look over the shoulders of capital and labor when they came to the bargaining table and its insistence that any agreement they reached would have to respect the national interest.

And in those seventy-two hours, new content and meaning were poured into that magnificent abstraction, "the Presidency," for the historically minded to argue about as long as men remained interested in the affairs of this republic.

A full and entirely accurate account of those seventy-two hours may never be written. The characters were many. They moved so fast that no one will be able to retrace all of what they did.

Understandably, industry participants—facing official investigation now—would not talk much. Nor were Government participants willing to tell all.

Nevertheless, a team of New York Times reporters undertook to piece the tale together while memories were fresh.

Here is what they learned:

Early on that afternoon of April 10, Roger Blough had met with his colleagues of United States Steel's executive committee in the board room on the twentieth floor at 71 Broadway, New York. Three of the twelve members were absent, but Leslie B. Worthington, president of the company, and Robert C. Tyson, chairman of the finance committee, were there.

Hints of Rise Given

For several months these men had been giving out hints, largely overlooked in Washington, that the company would have to raise prices to meet increasing costs.

The Kennedy Administration had striven last fall to prevent a steel price increase, and there had been no increase. It had pressed again for a modest wage contract this year, and a modest contract had been signed a few days earlier. The Administration expected no price increase now.

The company's executive committee reviewed the situation. The sales department had concurred in a recommendation to increase prices by 3½ per cent—about $6 on top of the going average of $170 a ton.

Mr. Blough had taken soundings within the company on the public relations aspects. Everyone realized that the move would not win any popularity prize, but the committee voted unanimously to go ahead.

With the decision made, Mr. Blough took a plane to Washington. Word was telephoned to the White House that he wanted to see the President and had something "important" to say about steel.

A few minutes after 5:45 the President received him in his oval office, motioned him to a seat on a sofa to his right and made himself comfortable in his rocking chair.

With little preliminary, Mr. Blough handed the President a four-page mimeographed press release that was about to be sent to newspaper offices in Pittsburgh and New York.

The President read:

"Pittsburgh, Pa., April 10—For the first time in nearly four years, United States Steel today announced an increase in the general level of its steel prices."

Mr. Kennedy raced through the announcement. Then he summoned Arthur J. Goldberg, the Secretary of Labor. Minutes later Mr. Goldberg reached the President's office from the Labor Department four blocks away.

Grimly, the President gave the paper to Mr. Goldberg and said it had been distributed to the press. Mr. Goldberg skimmed over it and asked Mr. Blough what was the point of the meeting, since the price decision had been made.

Mr. Blough replied that he thought he should personally inform the President as a matter of courtesy. Mr. Goldberg retorted it was hardly a courtesy to announce a decision and confront the President with an accomplished fact.

In the half-hour discussion that followed President Kennedy seems to have kept his temper. But Mr. Goldberg lectured Mr. Blough with some heat. The price increase, the Secretary said, would jeopardize the Government's entire economic policy. It would damage the interests of United States Steel itself. It would undercut responsible collective bargaining. Finally he said, the decision could be viewed only as a double-cross of the President because the company had given no hint of its intentions while the Administration was urging the United Steelworkers of America to moderate its wage demands.

Mr. Blough, a high school teacher turned lawyer and company executive, defended himself and the company in a quiet voice.

When he had gone President Kennedy called for the three members of his Council of Economic Advisers. Dr. Walter W. Heller, the chairman, a lean and scholarly looking man, came running from his office across the street. Dr. Kermit Gordon followed in three minutes. James Tobin, the third member, hurried back to his office later in the evening.

Into the President's office came Theodore C. Sorensen, the White House special counsel, Mr. O'Donnell and Andrew T. Hatcher, acting press secretary in the absence of Pierre Salinger, who was on vacation.

one hundred thirty-one

Now the President, who usually keeps his temper under rein, let go. He felt he had been double-crossed—deliberately. The office of the President had been affronted. The national interest had been flouted.

Bitterly, he recalled that:

"My father always told me that all business men were sons-of-bitches but I never believed it till now!"

It was clear that the Administration would fight. No one knew exactly what could be done, but from that moment the awesome power of the Federal Government began to move.

How It Developed

To understand the massive reaction of the Kennedy Administration, a word of background is necessary.

Nothing in the range of domestic economic policy had brought forth a greater effort by the Administration than the restraint it sought to impose on steel prices and wages.

Starting last May the Administration worked on the industry, publicly and privately, not to raise its prices when wages went up in the fall. And when the price line held, the Administration turned its efforts to getting an early and "non-inflationary" wage contract this year.

Above all, the Administration constantly tried to impress on both sides that the national interest was riding on their decisions. A price increase or an inflationary wage settlement, it argued, would set off a new wage-price spiral that would stunt economic growth, keep unemployment high, cut into export sales, weaken the dollar and further aggravate the outflow of gold.

On Friday and Saturday, April 6 and 7, the major steel companies had signed the new contract. President Kennedy had hailed it as "noninflationary." Privately, some steel leaders agreed with him.

Thus, the President confidently expected that the companies would not increase prices. And the standard had been set, he hoped, for other industries and unions.

This was the background against which the group in the President's office went to work.

By about 8 P.M. some decisions had been reached.

President Kennedy would deliver the first counter-attack at his news conference scheduled for 3:30 the following afternoon.

Messrs. Goldberg, Heller and Sorensen would gather material for the President's statement. Other material of a statistical nature would be prepared in a longer-range effort to prove the price increase was unjustified.

While the discussion was going on, the President called his brother, Robert F. Kennedy, the Attorney General; Secretary of Defense Robert S. McNamara, and the Secretary of the Treasury, Douglas Dillon, who had just arrived in Hobe Sound, Fla., for a short vacation.

At his home on Hillbrook Lane, Senator Estes Kefauver of Tennessee, chairman of the Senate Antitrust Subcommittee, was getting ready to go out for the evening. The phone rang. It was the President. Would Senator Kefauver publicly register "dismay" at the price increase and consider an investigation?

The Senator certainly would. He promised an investigation. So did the Justice Department.

In the President's office, meanwhile, there had been some talk of what could be done to keep other steel companies from raising prices. Most of the discussion centered on the economic rebuttal of the case made by United States Steel.

Mr. Goldberg and Dr. Heller decided to pool resources. Mr. Goldberg called Hyman L. Lewis, chief of the Office of Labor Economics of the Bureau of Labor Statistics, and asked him to assemble a crew.

Mr. Lewis reached three members of the bureau—Peter Henle, special assistant to the Commissioner of Labor Statistics; Arnold E. Chase, chief of the Division of Prices and Cost of Living, and Leon Greenberg, chief of the Productivity Division.

He told them what was wanted and asked them to go to Dr. Heller's office in the old State Department Building.

Dr. Heller who had been working on the problem in his office, hurried off after a few minutes to the German Ambassador's residence on Foxhall Road.

The Ambassador was giving a dinner, a black tie affair, in honor of Prof. Walter Hallstein, president of the European Common Market. The guests were well into the meal when Dr. Heller arrived, looking, as one of the guests remarked, like Banquo's ghost in a tuxedo.

White House Reception

Back at the White House the President had also changed to black tie. The members of Congress and their wives were coming to his annual reception at 9:45. Ruefully, the President recalled that the news of the Cuban disaster had arrived during his reception in 1961.

"I'll never hold another Congressional reception," he remarked.

But as he and Mrs. Kennedy received the leaders of Congress and their wives, he easily relaxed into small talk.

What did the men think, he asked, of the break with tradition by making this a black tie, instead of a white tie, affair? Republicans and Democrats unanimously favored the change. Many of the younger members of Congress, they pointed out, did not have a white tie and all that went with it.

With the party spread through three rooms, no one could tell how many times Mr. Kennedy slipped out to talk about steel. He stayed until 12:08 A.M. Then he retired.

By that time, the White House staff, the Council of Economic Advisers and the Departments of Labor, Justice, Defense, Commerce and The Treasury were all at work on the counterattack.

WEDNESDAY

Midnight had struck when Walter Heller, still in black tie, returned to his office from the German Embassy. With him, also in black tie, came another dinner guest, George W. Ball, Under Secretary of State.

Dr. Heller's two colleagues in the Council of Economic Advisers, Dr. Gordon and Dr. Tobin, were already there.

At about 2:45 A.M. the four men from the Bureau of Labor Statistics left the session. Their assignment from then on was to bring up to date a fact book on steel put out by the Eisenhower Administration two years ago.

The idea was to turn it into a kind of "white paper" that would show that the price increase was unjustified.

Toward 4 o'clock Dr. Heller and Dr. Tobin went home for two or three hours' sleep. Dr. Gordon lay down on the couch in his office for a couple of hours.

As the normal working day began, President Kennedy held a breakfast meeting at the White House with Vice President Johnson; Secretary of State Dean Rusk (who played no part in the steel crisis); Secretary Goldberg; Mr. Sorensen; Myer Feldman, Mr. Sorensen's deputy; Dr. Heller and Andrew Hatcher.

The meeting lasted an hour and forty-five minutes. Mr. Goldberg and Dr. Heller reported on the night's work. Mr. Sorensen was assigned to draft the

one hundred thirty-three

President's statement on steel for the news conference. Mr. Goldberg gave him a two-page report from the Bureau of Labor Statistics headed: "Change in Unit Employment Costs in the Steel Industry 1958 to 1961." It said in part:

> While employment costs per hour of all wage and salaried employes in the basic iron and steel industry rose from 1958 to 1961, there was an equivalent increase in output per man-hour.
>
> As a result, employment costs per unit of steel output in 1961 was essentially the same as in 1958.

The latter sentence was quoted that afternoon in the President's statement.

During the morning the President had called Secretary Dillon in Florida and discussed with him the Treasury's work on tax write-offs that would encourage investment in more modern plant and machinery. The two decided that the course would not be altered.

The President also telephoned Secretary of Commerce Luther H. Hodges, who was about to testify before a House Maritime subcommittee. After giving his testimony Secretary Hodges spent most of the day on the phone to business men around the country.

In Wall Street that morning United States Steel shares opened at 70¾, up 2¾ from the day before. But on Capitol Hill the company's stock was down.

Senator Mike Mansfield, the majority leader, called the price increase "unjustified." Speaker John W. McCormack said the company's action was "shocking," "arrogant," "irresponsible." Senator Hubert H. Humphrey, the Democratic whip, spoke of "an affront to the President."

Curb by Law Suggested

Senator Albert Gore of Tennessee suggested a law that would empower the courts to prohibit price increases in basic industries such as steel until there had been a "cooling-off period."

Representative Emanuel Celler of Brooklyn, chairman of the House Antitrust subcommittee, scheduled a broad investigation of the steel industry. So did Senator Kefauver.

The pressures on United States Steel were beginning to mount. But now some of the other titans of the industry began to fall in line behind Big Steel.

As the President came out of the White House shortly before noon to go to the airport where he was to welcome the Shah of Iran, he was shown a news bulletin. Bethlehem Steel, second in size only to United States Steel, had announced a price increase.

Others followed in short order—Republic, Jones and Laughlin, Youngstown and Wheeling. And Inland, Kaiser and Colorado Fuel & Iron said they were "studying" the situation.

When he faced the newsmen and television cameras at 3:30, President Kennedy spoke with cold fury. The price increase, he said, was a "wholly unjustifiable and irresponsible defiance of the public interest." The steel men had shown "utter contempt" for their fellow citizens.

He spoke approvingly of the proposed investigations. But what did he hope to accomplish that might still save the Administration's broad economic program?

In his conference statement the President had seemed to hold out no hope that the price increases could be rolled back. If the increases held, what imminent comfort could there be in possible antitrust decrees that would take three years to come from the courts?

Actually, the possibility of making United States Steel retract the increase had been considered early in the consultation.

Drs. Heller and Gordon, and possibly some of the other economists, had argued that the principal thrust of the Administration's effort should be to convince one or two significant producers to hold out. In a market such as steel, they said, the high-priced sellers would have to come down if the others did not go up.

This suggested a line of strategy that probably proved decisive.

Hold-outs Emerge

As one member of the Big Twelve after another raised prices, only Armco, Inland, Kaiser, C F & I and McLouth were holding the line. These five hold-outs represented 14 per cent of total industry capacity, or 17 per cent of the capacity of the Big Twelve.

Everything pointed to Inland as the key to the situation.

Inland Steel Corporation with headquarters in Chicago is a highly efficient producer. It could make a profit at lower prices than those of some of the bigger companies. And any company that sold in the Midwest, such as United States Steel, would feel Inland's price competition.

Moreover, there was a tradition of public service at Inland. Clarence B. Randall, a former chairman of the board, had served both the Eisenhower and Kennedy Administrations. (But he played no part in this crisis.)

Joseph Leopold Block, Inland's present chairman, who was in Japan at the moment, had been a member of President Kennedy's Labor-Management Advisory Committee.

At 7:45 that Wednesday morning, Philip D. Block Jr., vice chairman of Inland, was called to the telephone in his apartment at 1540 North Lake Shore Drive in Chicago.

"Hello, P. D.," said Edward Gudeman, Under Secretary of Commerce, a former schoolmate and friend of Mr. Block's, calling from Washington.

"What do you think of this price increase of United States Steel's?"

Mr. Block said he had been surprised.

"I didn't ask P. D. what Inland might do," said Mr. Gudeman several days later. "I didn't want them to feel that the Administration was putting them on the spot. I just wanted him to know how we felt and to ask his consideration."

Agree to Consider It

Inland officials said they had not been coaxed or threatened by any of the officials who called them.

The approach, which seems to have developed rather spontaneously in many of the calls that were made to business men, was to ask their opinion, state the Government's viewpoint, and leave it at that.

But there also were calls with a more pointed aim—to steel users, asking them to call their steel friends and perhaps even issue public statements.

Another call to Inland was made by Henry H. Fowler, Under Secretary of the Treasury and Acting Secretary in Mr. Dillon's absence.

After Mr. Kennedy's afternoon news conference Mr. Fowler called John F. Smith Jr., Inland's president. Like other Treasury officials who telephoned other business men, Mr. Fowler talked about the effect of a steel price increase on imports and exports and the further pressure it would place on the balance of payments.

A third call went to Inland that day. It was from Secretary Goldberg to Leigh B. Block, vice president for purchasing.

Both Inland and Government officials insist that there was no call from the White House or from any Government office to Joseph Block in Japan.

Though no concrete assurance was asked or volunteered in these conversations, the Administration gathered assurance that Inland would hold the line for at least another day or two.

Next came Armco, sixth largest in the nation. Walter Heller had a line into that company. So did others. Calls were made. And through these channels the Administration learned that Armco was holding off for the time being, but there would be no public announcement one way or the other.

Meanwhile, Mr. Gudeman had called a friend in the upper reaches of the Kaiser Company. Secretary McNamara had called a number of friends, one of them at Allegheny-Ludlum, a large manufacturer of stainless.

How many calls were made by President Kennedy himself cannot be told. But some time during all the activity he talked to Edgar Kaiser, chairman of Kaiser Steel, in California.

According to one official who was deeply involved in all this effort, the over-all objective was to line up companies representing 18 per cent of the nation's capacity. If this could be done, according to friendly sources in the steel industry, these companies with their lower prices soon would be doing 25 per cent of the business. Then Big Steel would have to yield.

Antitrust Line Pushed

Parallel with this "divide-and-conquer" maneuver, the effort moved forward on the antitrust line.

During the morning someone had spotted in the newspapers a statement attributed to Edmund F. Martin, president of Bethlehem Steel. Speaking to reporters on Tuesday after a stockholders' meeting in Wilmington, Del., Mr. Martin was quoted as having said:

> There shouldn't be any price rise. We shouldn't do anything to increase our costs if we are to survive. We have more competition both domestically and from foreign firms.

If Mr. Martin had opposed a price rise on Tuesday, before United States Steel announced its increase, and if Bethlehem raised its prices on Wednesday after that announcement, his statement might prove useful in antitrust proceedings. It could be used to support a Government argument that United States Steel, because of its bigness, exercised an undue influence over other steel producers.

F.B.I. Told to Check

At about 6 o'clock Wednesday evening, according to officials of the Justice Department, Attorney General Kennedy ordered the Federal Bureau of Investigation to find out exactly what Mr. Martin had said.

At about this same time, Paul Rand Dixon, chairman of the Federal Trade Commission, told reporters that his agency had begun an informal investigation to determine whether the steel companies had violated a consent decree of June 15, 1951.

That decree bound the industry to refrain from collusive price fixing or maintaining identical delivered prices. It provided penalties running up to $5,000 a day.

Meanwhile, more calls were going out from Washington.

The Democratic National Committee called many of the Democratic Governors and asked them to do two things:

First, to make statements supporting the President and, second, to ask steel producers in their states to hold the price line.

Among those called were David L. Lawrence of Pennsylvania, Richard J. Hughes of New Jersey and Edmund G. Brown of California. But the National Committee said nothing in its own name. The smell of "politics" was not to be allowed to contaminate the Administration's efforts.

A Morgan Man Called

Another call was made by Robert V. Roosa, an Under Secretary of the Treasury, to Henry Alexander, chairman of Morgan Guaranty Trust Company in New York. Morgan is represented on United States Steel's board of directors and is widely considered one of the most powerful influences within the company.

Thus by nightfall on Wednesday—twenty-four hours after Mr. Blough's call on the President—the Administration was pressing forward on four lines of action:

First, the rallying of public opinion behind the President and against the companies.

Second, divide-and-conquer operation within the steel industry.

Third, antitrust pressure from the Justice Department, the Federal Trade Commission, the Senate and the House.

Fourth, the mobilization of friendly forces within the business world to put additional pressure on the companies.

That night at the White House the Kennedys gave a state dinner for the visiting Shah and his Empress.

In a toast to his guests, President Kennedy, a man seemingly without a care in the world, observed that he and the Shah shared a common "burden." Each of them had made a visit to Paris and each of them might as well have stayed at home, for the Parisians had eyes only for their wives.

When the guests had gone, the President put in a call to Tucson, Ariz. It came through at 12:15 A.M.

THURSDAY

Archibald Cox, the Solicitor General, had left by plane on Wednesday afternoon for Tucson, where he was to make two speeches to the Arizona Bar.

On arriving at his hotel that night, he received a message to call the President. When he called he was asked what suggestions did he have for rolling back steel prices?

Mr. Cox had been chairman of the Wage Stabilization Board during the Korean War and had worked with young Senator Kennedy on statements about steel prices and strikes of the past.

After the call, Mr. Cox stayed up all night, thinking and making notes, mostly about legislation. From past experience Mr. Cox had concluded that the antitrust laws could not cope with the steel problem and that special legislation would be necessary.

Mr. Cox made his two speeches, flew back to Washington and stayed up most of that night working on the legislative draft.

But Mr. Cox was not the only one at work on the steel problem in the early hours of Thursday.

Awakened By F.B.I.

At 3 A.M. Lee Linder, a reporter in the Philadelphia bureau of the Associated Press, was awakened by a phone call. It was the F.B.I. At first Mr. Linder thought he was being fooled. Then he determined that the call was genuine. The agents asked him a question or two and then told him:

We are coming right out to see you.

Mr. Linder had been at the stockholders' meeting of Bethlehem Steel in Wilmington on Tuesday and had quoted Mr. Martin about the undesirability of a price increase. Bethlehem Steel later called the quotation incorrect.

The agents were checking on that quotation. Mr. Linder said later that he had given them the same report he had written for The Associated Press.

At 6:30 A.M. James L. Parks Jr. of The Wilmington Evening Journal arrived at

one hundred thirty-seven

his office. Two F.B.I. agents were waiting for him. He had talked to Mr. Martin after the meeting, together with Mr. Linder and John Lawrence of The Wall Street Journal. Later in the day the Federal agents interviewed Mr. Lawrence. This descent of the F.B.I. on the newsmen was the most criticized incident in the seventy-two frenzied hours.

Republicans, who had kept an embarrassed silence up to this point, pounced on this F.B.I. episode. Representative William E. Miller of upstate New York, chairman of the Republican National Committee, compared it to the "knock on the door" techniques of Hitler's Gestapo.

In Chicago, as the day progressed, Philip Block and two other high officials of Inland reached a decision: prices would not be raised. They called Joseph Block in Kyoto. He concurred and they agreed to call a directors' meeting to ratify their decision the next morning.

No announcement was to be made until the morning and no one in Washington was told.

Back in Washington, the President was holding an early meeting in the Cabinet Room at the White House. Present were:

Attorney General Kennedy; Secretaries McNamara, Goldberg, Hodges; Under Secretary of the Treasury Fowler; Mr. Dixon, chairman of the Federal Trade Commission; Dr. Heller and Mr. Sorensen.

Quick Rebuttal Planned

Roger Blough was scheduled to hold a televised news conference in New York at 3:30 that afternoon. The White House meeting decided that the Administration should put in a speedy rebuttal to his case for United States Steel.

Secretary Hodges had long-scheduled engagements that day in Philadelphia and New York. It was decided that he would hold a news conference in New York at 5 P.M. and try to rebut Mr. Blough point by point.

Meanwhile two of the most secret initiatives of the entire seventy-two hours had been set in motion.

Helps Friends to Meet

The first involved a newspaperman—Charles L. Bartlett, the Washington correspondent of The Chattanooga Times. All Mr. Bartlett would say later was:

I helped two friends get in touch with each other again.

One friend was President Kennedy—Mr. Bartlett and his wife are members of the Kennedy social set. The other friend was an officer of United States Steel. His identity has not been definitely established, but Mr. Bartlett knows Mr. Blough.

What came of this effort to reopen "diplomatic relations" is not known, although at least one Cabinet member thought it was useful. What came of the second secret initiative, however, can be reported.

At noon or earlier on Thursday President Kennedy phoned Clark Clifford, a Washington lawyer who had first come to national prominence as counsel for President Truman.

Secretary Goldberg, said the President, knew the officers of United States Steel very well and could, of course, talk to them on behalf of the Administration. But Mr. Goldberg, he went on, was known to the steel men mainly as an adversary.

For years he had been the counsel for the steel workers' union and one of their chief strategists in negotiations with the company. In view of this would Mr. Clifford, familiar as he was with the outlook of corporation executives through his law work, join Mr. Goldberg in speaking to United States Steel?

Supports President

Mr. Clifford agreed, flew to New York and met Mr. Blough. He presented himself as a friend of the disputants, but he made clear that he was in 100 per cent agreement with the President. His purpose, he said, was to see if a tragic mistake could be rectified. The mistake, he left no doubt, was on the company's side.

For fourteen months, he continued, President Kennedy and Mr. Goldberg had worked for healthy conditions in the steel industry. They had tried to create an atmosphere of cooperation in the hope of protecting the national interest. Now all this was gone.

The President, he went on, believed there had been a dozen or more occasions when the company's leaders could easily have told him that despite all he had done they might have to raise prices. But they never had told him. The President, to put it bluntly, felt double-crossed.

What Mr. Blough said in reply could not be learned. But he indicated at the end that he would welcome further talks and he hoped Mr. Clifford would participate in them. Mr. Clifford returned to Washington the same day.

Secretary Hodges, meanwhile, arrived at the University Club in New York at about 3:40, ten minutes after Mr. Blough had begun his news conference.

While Mr. Hodges shaved and changed his shirt, his assistant, William M. Ruder, tried to take notes on Mr. Blough's broadcast, but the static he heard sounded like the Grand Central shuttle.

The Blough news conference was held in the ground floor auditorium at 71 Broadway.

"Let me say respectfully," Mr. Blough began, "that we have no wish to add acrimony or misunderstanding."

On several occasions, he said, he had made it clear that United States Steel was in a cost-price torque that could not be tolerated forever, that a company without profits is a company that cannot modernize, and that the price increase would add "almost negligibly" to the cost of other products—$10.64 for the steel in a standard automobile, 3 cents for a toaster.

One question and answer in the fifty-eight-minute session caught the ears of people in Washington: Could United States Steel hold its new price if Armco and Inland stood pat?

"It would definitely affect us," conceded Mr. Blough. "I don't know how long we could maintain our position."

A half-hour after Mr. Blough finished, Secretary Hodges held his news conference in the Empire State Building.

But the words that probably hit Big Steel the hardest came that day from two Pennsylvania Republicans — Representatives William W. Scranton, the party's candidate for Governor, and James E. Van Zandt, the candidate for Senator.

"The increase at this time," they wired Mr. Blough, "is wrong—wrong for Pennsylvania, wrong for America, wrong for the free world. The increase surely will set off another round of inflation. It will hurt people most who can least afford to be hurt."

U.S. Serves Subpoenas

Meanwhile, Justice Department agents appeared at the headquarters of United States Steel, Bethlehem, Jones & Laughlin and other companies and served subpoenas for documents bearing on the price increase and other matters.

And at 7 P.M. Attorney General Kennedy announced that the Justice Department had ordered a grand jury investigation of the increase.

one hundred thirty-nine

By that time, President and Mrs. Kennedy were getting ready for another state dinner with the Shah and Empress—this time at the Iranian Embassy.

FRIDAY

The first big news of the day came from Kyoto, Japan. Joseph Block, Inland's chairman, had told a reporter for the Chicago Daily News:

> We do not feel that an advance in steel prices at this time would be in the national interest.

That news heartened the Administration but it did not stop planning or operations. Nor did Inland's official announcement from Chicago at 10:08 A.M., Washington time, that it would hold the price line.

At 10:15 Solicitor General Cox met in Mr. Sorensen's office with representatives of the Treasury, Commerce and Labor Departments, Budget Bureau and Council of Economic Advisers.

The discussion was on emergency wage-price legislation of three broad kinds:

First, ad hoc legislation limited to the current steel situation; second, permanent legislation imposing some mechanism on wages and prices in the steel industry alone, and third, permanent legislation for steel and other basic industries, setting up "fact-finding" procedures.

Defense Orders Shifted

At 11:45 Secretary McNamara said at his news conference that the Defense Department had ordered defense contractors to shift steel purchases to companies that had not raised prices. Later in the day the department awarded to the Lukens Steel Company, which had not raised prices, a contract for more than $5,000,000 worth of a special armor plate for Polaris-missile submarines.

At 12:15 President Kennedy and most of the Thursday group met again in the Cabinet Room. It was estimated at that time that the price line was being held on 16 per cent of the nation's steel capacity.

Inland had announced. Armco had decided to hold but not announce. Kaiser's announcement came in while the meeting was on. This might be enough to force the bigger companies down again, but the sentiment of the meeting was that the retreat would not come soon.

Accordingly, preparations continued for a long struggle. Lists of directors of the companies that were holding the line were distributed, and each man present was asked to call men he knew.

Notably absent from this meeting was Secretary Goldberg. He was on his way to New York with Mr. Clifford in a Military Air Transport plane.

A secret rendezvous had been arranged with Mr. Blough and some of the other leaders of United States Steel at the Carlyle Hotel.

At this meeting, as in Mr. Clifford's talk with Mr. Blough on the previous day, no demands or threats or promises came from the Government side.

Finds Outlook 'Abysmal'

The discussion seems to have been a general one about what lay ahead. The outlook, said Mr. Clifford, was "abysmal."

United States Steel, he contended, had failed to weigh the consequences of its action. If it held this position, its interest and those of the industry would inevitably be damaged, and the nation as a whole would suffer.

While the talk was going on, Mr. Blough was called to the phone. Then Mr. Goldberg was called. Each received the same message. Bethlehem Steel had rescinded the price increase—the news had come through at 3:20 P.M.

President Kennedy heard the news while flying to Norfolk for a weekend with the fleet. It was unexpected.

The Administration had made no special effort with Bethlehem. To this day, officials here are uncertain what did it.

Among other things, Bethlehem's officials were struck by the Inland and Kaiser announcement that morning. Inland posed direct competition to Bethlehem's sales in the Midwest—the largest steel market—and Kaiser posed it on the West Coast.

Further, special questions were raised by the Pentagon's order to defense industries to shift their steel buying to mills that did not raise prices. What did this mean for Bethlehem's vast operations as a ship builder?

Whatever the compelling factors were, Bethlehem's decision brought the end of the battle clearly in sight. The competitive situation was such that United States Steel's executive committee was not called into session to reverse its action of the previous Tuesday. The company's officers acted on their own.

The big capitulation came at 5:28. Mrs. Barbara Gamarekian, a secretary in the White House press office, was checking the Associated Press news ticker. And there was the announcement—United States Steel had pulled back the price increase.

Mrs. Gamarekian tore it off and ran into the office of Mr. Sorensen, who was on the phone to the acting press secretary, Mr. Hatcher, in Norfolk.

"Well," Mr. Sorensen was saying, "I guess there isn't anything new."

Mrs. Gamarekian put the news bulletin under his eye.

"Wait a minute!" shouted Mr. Sorensen.

Mr. Hatcher gave the news to the President as he came off the nuclear submarine, Thomas A. Edison, in Norfolk.

It was just seventy-two hours since Roger Blough had dropped in on Mr. Kennedy.

one hundred forty-one

EXHIBIT F   *Excerpts from Editorials and News Columnists Comments*

New York Times—The decision of United States Steel Corporation and other major producers to raise steel prices by $6 a ton is, as President Kennedy said yesterday, in "irresponsible defiance of the nation's interests." It imperils the economic welfare of the country by inviting a new tidal wave of inflation after four years of general price stability. It aggravates steel's competitive difficulties in domestic and foreign markets . . .

Wall Street Journal—It is certainly quite possible that the company's officials misgauged the market and will find that they cannot sell their steel at higher prices. Steel prices are under heavy competitive pressures from domestic production and foreign imports, not to mention competition from other construction materials. That decision will be made by the only instrument capable of measuring it, the competitive forces in the market place.

New York Daily News—The full facts, we feel, may clarify this story considerably. A steel company, even as a steel worker, has to make money if it is to survive. Despite soaring costs of doing business, our steel producers haven't boosted the price of their metal since a $4.50 per ton increase in 1958. After Washington's anti-business block has had its say, we suggest that all heads of this fundamental U. S. industry lay their facts plainly on the line for evaluation.

New York Herald Tribune—The steelmakers, in short, have been caught in a cost squeeze of which labor costs are only a part . . . . . It is not enough that wage costs per ton not increase. If the price line is to be held, they have to be reduced sufficiently to balance the rise in non-labor costs—specifically, to balance the increased capital per worker required for increased productivity, on which in turn depends the promise of a more bountiful economy for all. And the money for this investment can only come, ultimately, from healthy profits.

New York Mirror—The action of the United States Steel Corporation in raising prices, and President Kennedy's objection thereto, require more than superficial shouting about inflation. Is the price hike "wholly unjustifiable and irresponsible defiance?" That remains to be demonstrated. It is unrealistic to blame the companies for meeting the new costs with new prices.

Newark (New Jersey) Evening News—Ramifications of United States Steel's price increase are almost limitless, raising the spectre of higher prices along the whole economic front. Whatever legitimate claims steel may have for an increase were certainly prejudiced by its headlong rush to raise prices weeks before the new (labor) contract became effective. If the price increase becomes general, it would scarcely make United States manufacturers competitive with foreign industry.

Christian Science Monitor (Boston)—Insofar as the American steel industry follows the price-rise lead, it will have engaged in a damaging job of public relations. The "cold fury" of President Kennedy at the move may partly be written off as political disappointment. Yet his indictment has certainly been heard by a large audience.

Buffalo Evening News—While the industry can be harshly criticized for inviting the Presidential denunciation by the unexpectedness of its act, we will do well to bear in mind that there is a far more fundamental issue involved here, and it is this: Can the steelmakers of this country still fix the price at which they will sell their product, or must they yield to government fiat?

Baltimore Sun—His [President Kennedy's] anger is justified. Implicit in last week's settlement as understood by the White House and as understood by the public, was price stability in steel. The steel industry as a whole, we now learn, privately understood it otherwise.

Washington Post—The era of good feeling between government and business that seemed to be at hand may yield to a very different climate.

Washington Evening Star—If Big Steel's price hike was the opening gun in a planned war with the Kennedy Administration, as some people believe, the battle has been fully joined. And nothing could be more certain than the ultimate clobbering of the industry. It will adversely affect our ability to compete in foreign markets and to meet competition from abroad in our domestic markets. It will make it harder to stem the outflow of gold which is threatening the value of the dollar.

Louisville Times—The increase does not necessarily mean inflation, but is a powerful threat. Only those with access to the complete cost structure of the steel industry could make a confident judgment about the justification for the price rise.

Atlanta Constitution—President Kennedy's bitter denunciation will not be lost on the American public. As the President said, business men, farmers, home owners, all consumers will share in this cost, which will open another Pandora's box of wage-price hikes.

Cleveland Plain Dealer—The President, his Council of Economic Advisers, and Labor Secretary Arthur J. Goldberg virtually asked for what happened when they assured labor that there was plenty of leeway for wage increases "within the limits of productivity." They encouraged labor to grab all the benefits of productivity without regard to the investors who put up the money to buy the modern equipment which made increased productivity possible.

Columbus (Ohio) Dispatch—It would appear Mr. Kennedy, at his press conference, left the impression that the steel industry, in some manner or other, had broken faith on the recently concluded contract agreement. Yet Mr. Kennedy says he had no commitment.

Hammond (Indiana) Times—Certainly U. S. Steel listed some compelling reasons for raising its prices. It pointed out that the corporation, the largest steel producer in the nation, last adjusted its overall prices back in the summer of 1958. Since that time it has absorbed the costs of four increases in wages and benefits paid to steelworkers, and a fifth is on the way under the recent settlement.... At the same time, it has kept its stockholders on the same dividend ration for more than five years.

Indianapolis Star—Now, if it would be bad for the steel companies to agree among themselves to raise prices—and it would be bad—then it is equally bad for them to agree not to raise prices. As a matter of fact, a group of private firms can get into just as bad trouble with the Antitrust Division through conspiracy to set low prices as through conspiracy to fix a high price. If the President was party to an understanding that steel prices would remain fixed at previous levels, he was playing fast and loose with the principles of the law he now seems so eager to invoke.

Chicago Sun-Times—It is politically popular to decry corporation price increases but not worker cost increases. A reading of the reasons for the price boost certainly show that it is not an arbitrary or greedy action. Employment costs have been raised seven times in more than five years, but dividend rates have not been raised at all.

Chicago Daily News—The outraged cries in Congress leave no doubt that the price hike was poorly timed, although it would be hard to say just when would

one hundred forty-three

have been the expedient time. To the dispassionate observer, however, the question posed is whether these signs do not point in the direction of eventual nationalization of the steel industry.

St. Louis Globe-Democrat—In the fact of other big steel companies' refusal to go along with a $6 a ton increase, United States Steel rescinded the price boost it announced . . . . Whether steel can continue to hold the price line, without coming apart at the fiscal seams, is another question. It has absorbed four wage and fringe increases since 1958 without upping prices . . . . It is most pertinent also to ask why Mr. Kennedy did not bring his great influence to bear against the outrageous demands of electrical workers in New York, who sought and got a five-hour day and a 56-cent wage rise . . . . We hope inflation can now be stemmed in steel, our basic industry. It obviously can't if steady wage boosts continue. No efforts will halt inflation unless they are directed toward holding the line on pay costs as well as prices.

St. Louis Post Dispatch—If the increase is justified by higher employment costs, as industry leaders contend, then it is the bequest of the January, 1960, wage settlement, which was engineered by Vice President Richard M. Nixon. It looks very much as if the steel masters used the President and his Secretary of Labor for beating down wage demands prior to a price decision they had in mind all along.

Omaha World-Herald—We cannot go along with the loud voices in Washington which are calling it "shocking and irresponsible," while they place all the blame at management's door and absolve the steel union totally.

Milwaukee Journal—President Kennedy chose the right words with which to denounce the increase in the price of steel. A quick admission of their error by the steel executives and a swift restoration of the former price level would be a truly responsible act.

Minneapolis Tribune—Hopes were pitted against realities in steel, realities won. The disappointment and wails of anguish are understandable, although the reaction of President Kennedy seemed unnecessarily bitter. Can the industry remain competitive? That question throws another dark cloud over a steel scene newly shrouded in disappointments.

Oklahoma City Oklahoman—In a free economy, management must be the final judge of its own necessities. If politicians insist on inflating industry's costs they must acknowledge the consequences in terms of rising prices. Otherwise they threaten to destroy the free enterprise system.

Tulsa Tribune—Perhaps the whole dispute is getting academic. Maybe both steel labor and the steel industry are about to be punished by foreign competition on a scale never dreamed of. Then, if American industry is profiteering, it will destroy itself.

Denver Post—As Mr. Kennedy said, the steel men were asked, like all other Americans, to consider what they could do for their country in a time of trouble. In the dangerous price increases they instituted . . . the nation has had their answer.

Los Angeles Times—Mr. Kennedy descended to a level of demagoguery at his press conference that has rarely been achieved by Senator Kefauver. If Mr. Kennedy had addressed the union with the same scathing rhetoric when the members got their 10 cents an hour, his position might seem more reasonable.

Oakland (California) Tribune—In fact, the missing element in President Kennedy's panacea for all is the fact that profits are gone, missing, done for . . . . Many economists are agreed that the missing elements in the Kennedy program is the fact that profits are being undercut by political uncertainties. Notice the stock market . . . . . We don't state categorically that this may be the reason for the President's blast against the steel companies, but it is always nice to have someone to blame. Particularly during a congressional election year.

Richmond (California) Independent—President Kennedy, a Harvard man, should know enough simple economics to be aware that as long as the government is spending more money than it is taking in, as long as the men in Washington refuse to balance the budget, the dollar is being diluted and inflationary pressures are pushing prices upward in a never ending spiral. We think President Kennedy knows this, but it would be bad politics to lay the blame on his own doorstep. He needed a whipping boy.

Portland Oregonian—President Kennedy says he did not set the price of steel. The truth is, of course, that had Mr. Kennedy not launched his blockbuster attack the major steel companies would have followed U. S. Steel in making the moderate price increases all say are necessary for plant investment in a competitive world. In that event, U. S. Steel would not have been forced by competition at home—competition created by the Administration—to knuckle down. President Kennedy reset the steel price as surely as if he had done it by executive order . . . . Our guess is that the steel formula, enforced by President Kennedy in what can only be described as a misuse of executive power, will govern only in steel.

Boise (Idaho) Daily Statesman—There's more than one way of looking at this steel price boost . . . . The approach which fair play demands . . . undertakes, with as much impartiality as may be, to pick up the pieces that are minimized or entirely ignored in the argument . . . . In this connection it cannot escape attention that there's no more basis for the claim that the lately negotiated steel labor contract is "non-inflationary" than the mere assertion by the President himself that is echoed by his claque including Idaho's Senator Church, who knows less about steel or economics than he does about Sanskrit.

Barron's Magazine—In his bitter outburst of last April, President Kennedy pointedly spoke of the handsome earnings and dividends enjoyed by the steel industry and its shareholders. Subsequent events suggest that his figures, and the judgments to which they contributed, were equally faulty. True, what has been done cannot be undone. However, it's high time that somebody broke the news to such anti-business stalwarts as Senator Estes Kefauver. Steel is in trouble. Surely the least Washington can do is get off its back.

Michael O'Neill, Washington Column in *The New York Daily News*—Most Americans are under the impression that U. S. machinery and equipment are the most modern in the world. But this is simply not true in the case of many industries—textiles, for example—and the result is that many foreign manufacturers are able to produce much more efficiently than U. S. firms. The "crisis of obsolescence" is particularly acute in the steel industry. . . . The serious lag in modernization is greatly aggravated by the fact that foreign competitors have been pouring relatively larger sums into updating their equipment. European firms, for instance, are spending twice as much proportionately as U. S. companies.

David Lawrence, Washington Column in *The New York Herald Tribune*—A new era in American history—a declaration of war by the government on the profit system as it functions under private capitalism—has been ushered in by President Kennedy.

Donald I. Rogers, Financial/business Column in the *Herald Tribune*—It's pretty easy for anyone but a blind man to tell a car from a horse. Mr. Kennedy's

one hundred forty-five

fury at the steel industry for its inflationary price increase is the reaction of a man who is so mad he can't tell the difference. The steel price increase is the result of the inflation that has already occurred and has been masked in recent months, rather than the cause of any future inflation.

Arthur Krock, Washington Column in *The New York Times*—The timing and all the other imperious circumstances attending the announcement of an immediate price rise by United States Steel Corporation made it a declaration of war . . . . And, though the industry's policy-makers must have anticipated a belligerent reaction, the intensity of the President's withering counter-fire may have surprised them. . . . Except for the mild demurrer issued by their Chief negotiator, R. Conrad Cooper, on the signing of the new contract with the steelworkers (that the settlement cost to the companies did not fit entirely within the framework of "anticipated gains in productive efficiency") no steel executive had given the President any reason to expect that his own optimistic appraisal would be discredited and in less than two weeks, by industry action.

Walter Lippmann, Syndicated Column of Comment on Public Affairs—Coming so soon after the wage settlement, the decision of United States Steel Corporation to raise its prices about $6 a ton has the look of a defiant repudiation of the basic understanding. For it rested on the assumption that there would be no rise in steel prices, if the new benefits for labor were well within the current increase in productivity . . . . This assumption has now been shattered, and it must be said shattered rudely by the decision of the company made without previous notice to or consultation with anyone speaking for the national interest.

Peter Edson, Washington Column in *The New York World-Telegram & Sun*—If the union had known that management intended to raise steel prices, the union would have demanded more.

EXHIBIT G *Statements by Leaders in Public Office, Politics, Labor, and Business*

David J. McDonald, President of the United Steelworkers of America, issued the following statement from his headquarters in Pittsburgh the night of April 10, 1962:

I am surprised, troubled, and concerned by the announcement of United States Steel that steel prices are being increased by $6 per ton.

The Steelworkers Union has never bargained with the steel companies about prices and did not do so this year.

We had no understanding and no discussion with the company on prices. I was surprised by the announcement, nevertheless, because the fact is that, as the President of the United States has said, our settlement was entirely non-inflationary and well within the current increase in steel productivity.

I am troubled by United States Steel's attempt to place the blame for this price increase upon settlements which have been made in the past by the United Steelworkers.

The fact is that since 1958, the date of the industry's last price increase, the labor cost of producing a ton of steel has gone down, not up, despite the increase and wages and other benefits which have been negotiated in that period.

I am concerned by this announcement because of the possible effect of the price increase on the nation's economy and position in the world.

George Meany, President of the American Federation of Labor and Congress of Industrial Organizations, attacked the price increase as "price gouging," and said the steel labor settlement was "non-inflationary."

Walter Reuther, President of the United Automobile Workers of America and Vice President of the AFL/CIO, supported President Kennedy's position, in letters to the President and a group of Senators and Representatives, and proposed establishment of a board to hold hearings before a price increase such as that in steel could take effect.

George Romney, former automobile company executive and then Republican candidate for Governor of Michigan, said "the solution is not destruction of our competitive system through substitution of government power or control. The solution is the strengthening of our competitive principles on the basis of proven political and economic concepts."

Secretary of Commerce *Luther Hodges* (in a news conference on April 12, 1962, that followed the news conference held by the Board Chairman of U. S. Steel) said the price increase would "reduce the competitiveness in world markets of many other products than steel itself... " and "could lead to a change in the economic philosophy and program" of the Kennedy Administration.

Senator Mike Mansfield (Democrat, Montana), majority leader in the United States Senate, called the increase "unjustified." . . . The Justice Department will take a "more than ordinary interest" in the situation . . . . "The time for talking is past. The time for action has come."

Representative John W. McCormack (Democrat, Massachusetts), Speaker of the House, called U. S. Steel's action "shocking ... arrogant ... irresponsible."

Senator Everett McKinley Dirksen (Republican, Illinois), minority leader in the Senate, said "The steel controversy must be determined by the facts. We would suggest that Mr. Kennedy is looking in the wrong place for the basic cause of inflation," which Senator Dirksen described as "excessive Government spending."

Representative Carl Albert (Democrat, Oklahoma), majority whip in the House, commented: "How can anyone justify a raise in steel prices when the

one hundred forty-seven

labor costs of the steel industry are the same today as they were in 1958 . . . when the demand for steel has fallen."

Senator Homer Capehart (Republican, Indiana), said the price increases are "a result of the President's trying to dictate American business. When you get a President who insists on being a czar, and running industry, this is what you get."

Senator Barry Goldwater (Republican, Arizona), said President Kennedy's attack on the steel companies was "something you'd expect in a police state."

Senator Jacob Javits (Republican, New York) criticized the Administration for using the Department of Justice and the Federal Bureau of Investigation "in a punitive way."

Representatives James E. Van Zandt and William W. Scranton (Republicans, Pennsylvania), who were then Republican organization candidates for, respectively, the senatorial and gubernatorial nominations in that state [this was before the party primary and their nominations], telegraphed the Chairman of the Board of U. S. Steel asking that the price increase be reconsidered, on the ground that the "increase this time is wrong—wrong for Pennsylvania, wrong for America, wrong for the free world."

Senator John G. Tower (Republican, Texas) said it was "sheer nonsense for the Administration to negotiate the steelworkers' fringe benefit increase, and then announce it would not be reflected in consumer costs . . . . President Kennedy's statement did not conform to his "demonstrated attitude" on Federal costs and is "inconsistent with his reaction to the New York electrical workers' successful demand for a 25-hour week and higher take-home pay."

Representative Charles A. Vanik (Democrat, Ohio) called U. S. Steel's action "a breach of faith."

Representative Ed Edmonson (Democrat, Oklahoma) said the price increase was "a very depressing development for the stability of America's economy."

Representative Edward P. Boland (Democrat, Massachusetts), said the pattern has been set and "everyone is now on the price rise escalator."

Representative August E. Johansen (Republican, Michigan) said "I am not at all favorably impressed with government by Presidential anger. I trust that Mr. Kennedy's televised outburst over the steel price increase will not establish a precedent."

Senator Stuart Symington (Republican, Missouri) said "the implication in some of the press . . . was that the regret expressed by the President about the decision on the part of some of the large companies to raise prices was an 'antibusiness' move. Nothing could be farther from the truth. The fact is that small-, medium-, and large-size businesses all over the country will now be faced with an increase in cost . . . . "

Representative Ralph Harvey (Republican, Indiana) said "The subject of the controversy is not new; Presidents in the past have tried to use their office to deal with wage negotiations within the steel industry. The impact of this effort will be felt for years to come."

Representative E. Y. Berry (Republican, South Dakota) said "the drift toward Government taking more and more power into its hands has been caught in the tidal sweep and is being washed directly toward centralization of authority in Washington. The tidal wave was further clearly evidenced by [the] thrust against steel."

Representative Carleton J. King (Republican, New York) said "The President and his advisers appear to have substituted their emotions for intelligence in the bitter clash with the steel industry and might well have waited to get all the facts in the matter before shooting from the hip . . . . Just how far the Administration's criticism, if not its vindictiveness, will go and how lethal it will be will not be known for several months."

Representative Jeffery Cohelan (Democrat, California) said "the welcome decision" of the steel companies to rescind their price increases is not only a victory for the President in his efforts to maintain a stable economy, but a 'victory for the self-regulatory forces of our free enterprise system' [quoted by Mr. Cohelan, from a New York Times editorial]."

The Joint Senate-House Republican Leadership of Congress issued a formal statement on April 19, 1962, asserting that nine specific actions taken by the Administration after U. S. Steel's price increase was announced, "amount to a display of naked political power never seen before in this nation" . . . that the price controversy raised a fundamental issue: "Should a President of the United States use the enormous powers of the Federal Government to blackjack any segment of our free society into line with his personal judgment without regard to law."

John M. Bailey, Democratic National Chairman, replied to the Republican statement, saying its criticisms of President Kennedy's action was "arrant nonsense" . . . a "silly and specious bit of partisan exaggeration . . . so clumsy it is sure to backfire on the Republican Party."

E. B. Germany, President of Lone Star Steel Company (one of the companies that did not raise its prices in mid-April of 1962), said: "Using tactics never before assumed by an Administration in peacetime, the President assumed the role of prosecutor, judge, and jury, to bring unprecedented influences to bear upon the steel industry. He assumed dictatorial powers when he directed that Government purchases of steel be made from companies which did not increase prices."

H. C. Ramsey, Chairman of the Worthington Corporation, makers of pumps, generators, and air conditioning equipment, said: "The steel industry and the capital goods industries need higher prices. We've struggled to raise our prices to a more realistic level, so that we can make a fair margin of profit . . . . Basically, the success of a company depends on its ability to get adequate prices for its products. The steel industry has absorbed wage increase after increase. The profit squeeze that was plaguing them is troubling us, too."

Thomas F. Patton, President of Republic Steel Corporation, said at his company's annual meeting of stockholders (May 9, 1962), that Republic "still needs a price increase . . . . We recognize that any citizen, including a government official, is entitled to try to persuade a company to forego a particular course of action, such as raising its prices, even though the company has a clear legal right to take such action. We insist, however, that such efforts at persuasion by a government official should not be accompanied by grand jury actions, visits from agents of the Federal Bureau of Investigation, the cancelling of government steel orders, and investigation by committees of Congress."

one hundred forty-nine

EXHIBIT H   *Press Release—United States Steel Corporation*
April 12, 1962

Statement By ROGER M. BLOUGH, Chairman, Board of Directors
UNITED STATES STEEL CORPORATION

When the President of the United States speaks as he did yesterday regarding our Corporation and its cost-price problems, I am sure a response is indicated and desirable.

Let me say respectfully that we have no wish to add to acrimony or to misunderstanding. We do not question the sincerity of anyone who disagrees with the action we have taken. Neither do we believe that anyone can properly assume that we are less deeply concerned with the welfare, the strength and the vitality of this nation than are those who have criticized our action.

As employees and stockholders, we along with thousands of other employees and stockholders, both union and non-union—must discharge faithfully our responsibilities to United States Steel; but as citizens we must also discharge fully our responsibilities to the nation. The action we have taken is designed to meet both of those responsibilities.

One of the nation's most valuable and indispensable physical assets is its productive machinery and equipment because its strength depends upon that. I among others share the responsibility of keeping a portion of that plant, machinery and equipment in good working order. To do that our company, like every other employer, must be profitable. The profits which any company has left over after paying its employees, its other expenses, the tax collector, and its stockholders for the use of their resources, are the main source of the plants and equipment that provide the work that thousands of workers now have. Had it not been for those profits in the past, the millions with jobs in many varieties of business would not have those jobs.

But that machinery and equipment must be kept up to date or no sales will be made, no work provided, no taxes available and our international competitive position, our balance of payments, our gold reserves and our national growth will seriously suffer.

None of us is unaware of the serious national problems and no one is unsympathetic to those in the executive branch of government attempting to conduct the affairs of the nation nationally and internationally. But certainly more rapid equipment modernization is one of the nation's basic problems as outlined by Secretary of the Treasury Douglas Dillon.

Speaking before the American Bankers Association last October, he said:

> More rapid equipment modernization by industry is vital to the success of our efforts to remain competitive in world markets and to achieve the rate of growth needed to assure us prosperity and reasonably full employment. I think it highly significant that all the industrial countries of Western Europe—except Belgium and the United Kingdom—are now devoting twice as much of their gross national product to purchases of industrial equipment as are we in the United States. And Belgium and the United Kingdom—the two European countries whose economic growth has lagged in comparison with the rest of Western Europe—are devoting half again as much of their GNP to the purchase of equipment as are we.

What this all means is that as a nation we keep ahead in the race among nations through machinery and equipment, through good productive facilities, through jobs that are vitally linked to the industrial stream. Surely our workmen are as good as any in the world, but they must have the tools with which to compete. In other words, we compete as a company, as an industry and as a nation with better costs and better ways of production.

Proper pricing is certainly part of that picture and that is what is involved here, however it may be portrayed. For each individual company in our competitive society has a responsibility to the public as well as to its employees and its stockholders to do the things that are necessary pricewise, however unpopular that may be at times, to keep in the competitive race.

Now, may I say several things with respect to any misunderstandings that have been talked about.

First, the President said, when questioned regarding any understanding not to increase prices, "We did not ask either side to give us any assurance, because there is a very proper limitation to the power of the government in this free economy." Both aspects of this statement are quite right. No assurances were asked and none were given regarding price action so far as I am concerned or any other individual connected with our Corporation. Furthermore, at least in my opinion, it would not have been proper for us under those circumstances to have had any understandings with anyone regarding price.

Second, I have said a number of times over the past months that the cost-price relationship in our company needed to be remedied. As recently as February 16, while the labor negotiations were going on, I referred, in an interview, to the steadily rising costs which we have experienced and I said "And you're asking me how long that can continue to increase and how long it can be borne without some kind of a remedy? I would give you the answer that it is not reasonable to think of it as continuing. In other words, even now there should be a remedy. If any additional cost occurs, the necessity for the remedy becomes even greater." This very real problem has been discussed in recent months with a number of individuals in Washington and I am at a loss to know why anyone concerned with the situation would be unaware of the serious deterioration in the cost-price relationship.

In this connection President Kennedy, in his letter of last September 6th addressed to executives of the steel companies, said: "I recognize too that the steel industry by absorbing increases in employment costs since 1958, has demonstrated a will to halt the price-wage spiral in steel." I am sure that anyone reading the reply that I made on September 13th to that letter could not infer any commitment of any kind to do other than to act in the light of all competitive factors.

Third, it is useful to repeat here that hourly employment costs have increased since 1958, by a total of about 12 per cent and that other costs have risen too. The net cost situation, taking into account employment and other costs, has risen about 6 per cent since 1958. All this is without regard to the new labor contract. When costs keep moving upward and prices remain substantially unchanged for four years, the need for some improvement in the cost-price relationship should be apparent.

Fourth, in view of the cost increases that have occurred, the thought that it costs no more to make steel today than it did in 1958 is quite difficult to accept. For U. S. Steel, costs since 1958 have gone up far more than the announced price increase of yesterday.

Fifth, higher costs at the same selling prices obviously mean lower profits. Our own profits of 5.7 per cent on sales in 1961 were the lowest since 1952.

Sixth, the increase of three-tenths of a cent per pound in the price of steel adds almost negligibly to the cost of the steel which goes into the familiar, every-day products that we use. Here, for example, is the amount by which this price change would increase the cost of steel for the following items:

Automobiles
    Standard size $10.64
    Intermediate size   8.33
    Compact size   6.83
Toaster   .03
Washing Machine—wringer type   .35
Domestic Gas Range—4-burner   .70
Refrigerator—7.7 cu. ft.   .65
303-size Food Can ·   5 one-hundredths of a cent

Seventh, it must be remembered that the process by which the human needs of people are met, and the process by which jobs are created, involve importantly the role of the investors. Only when people save and invest their money in tools of production can a new productive job be brought into existence; so our nation cannot afford to forget its obligation to these investors. Nor can we, in United States Steel—who are responsible to more than 325,000 stockholders —forget the many Americans who have a stake in our enterprise, directly or indirectly.

Over half of our shares are held by individuals in all walks of life; and no one of these individuals owns as much as two-tenths of one per cent of the total stock. Most of the rest is held by pension funds, insurance companies, charitable and educational institutions, investment companies and trustees representing the direct or indirect ownership of large numbers of people in America.

I have touched upon a few matters here in the hope that those who are so seriously concerned with these things—and the public at large—will recognize that there was nothing irresponsible about the action we have taken. My hope is that this discussion of our responsibilities, as we see them, will lead to a greater understanding, and a more thoughtful appraisal, of the reasons for that action.

* * * * * *

Questions and Answers

Roger M. Blough

News Conference, April 12, 1962

MR. CRONKITE: Mr. Blough, you say that no commitment was asked or given during the wage negotiations regarding a price increase. And yet, at your joint news conference with Mr. McDonald, he speaking first, mentioned the non-inflationary nature of the agreement, and the newspapers, radio and television played that as the biggest feature of the agreement.

I am wondering if you can tell us why there was no denial at that time or in this week that is past, on your part, that an agreement that an increase was intended.

MR. BLOUGH: Well, Mr. Cronkite, I do not think that Mr. McDonald was talking about the agreement that had just been concluded; that he was referring in any way, shape, or form to the price of steel.

There wasn't any occasion, therefore, to make any denial. You might have a slight difference of opinion with Mr. McDonald, as to whether a 2½ percent cost increase, which is required by the new agreement, is inflationary or non-inflationary. That gets you into quite a discussion.

In our opinion, the output per man hour of steel that we have experienced since 1940 is about 1.7 per cent, which is considerably below the 2½ per cent required by the contract.

BILL RAPLEY: Time Magazine: In the face of the competition that steel is facing, not only from foreign competition, but from other commodities in this country, commodities which are competing with you on price, how can you justify to the stockholders the increase in price in steel at this time?

MR. BLOUGH: I thought that my statement pretty well explained that.

MR. RAPLEY: That explains the increase in profits, sir, that you seek to achieve. But you do not run a risk of a distinct loss in volume?

MR. BLOUGH: In my statement I said you become non-competitive. Your costs are higher, if you have insufficient prices to provide the machinery and equipment that is necessary to remain competitive.

Our problem with respect to foreign competition is certainly a very serious one. There is no question that the plants and equipment abroad are, in many cases, the equal of ours and that we are having quite a difficult time from the standpoint of foreign competition. There is no question about that.

But if you do not provide the new plant and equipment that is necessary to keep up with that foreign competition, you are going to fall behind in the race much farther than you have. Now, I pointed out in my statement that Mr. Dillon said that, in many of those nations, they were providing two times the amount of their gross national product that we are providing in this nation, to provide the plant and equipment that is necessary to compete.

That is the serious problem that our nation faces, and that is the reason you have got to have sufficient prices to provide some profit which can, in turn, be plowed back into that machinery and equipment.

DAVE DUGAN: CBS News: Mr. Blough, you say in your statement that no assurances were asked by the federal government, and none was given by U.S. Steel, regarding price action. How, then, do you account for the intense reaction by President Kennedy to your price increase?

MR. BLOUGH: I quoted from President Kennedy's statement in response to a question in which he said that he hadn't asked for any assurance from us. I think that is a sufficient support for my statement.

TOM PETIT: NBC News: You here today, as I understand it, are defying the

one hundred fifty-three

President of the United States in his request yesterday. What are your views in general on the White House role in the steel labor negotiations, and the price dispute? Do you think the President has been acting in the public interest?

MR. BLOUGH: Well, now, first let me correct your preamble to your question. I am not here today in any sense defying anyone. I would like to make that perfectly clear. I am here today explaining the cost-price situation in which we found ourselves, and why we took the price action.

With respect to the action of the White House in connection with labor matters, I would say that I think that is a matter that can better be handled by the White House than by me. I have no criticism. I do believe that when the air clears a little bit, I think we will all realize that this type of, shall I say, assistance has some limitations.

RAD MCLANE: Associated: Have you been served with a subpoena to appear before a U.S. Grand Jury investigating matters relating to steel?

MR. BLOUGH: Would you ask the question again? Repeat the question, please.

RAD MCLANE: Have you, sir, been served with a subpoena to appear before a federal Grand Jury investigating matters relating to steel?

MR. BLOUGH: My understanding is that the company has been served with a subpoena to produce some papers, but I haven't personally been served with any subpoena to appear before the Grand Jury.

LARRY NATHAN: WMCA News: If some price rise was a necessity, why wasn't the government and the union made aware of this need?

MR. BLOUGH: Well, I tried to make perfectly clear in my opening statement that I personally have been talking about this sort of thing for a long time. I wonder if you will recall that last September there appeared to be a thought on the part of some people that some company in the steel industry might increase its prices.

You recall what happened there. Letters were exchanged between the president and the steel companies, with respect to that matter. Without indicating the effect of the exchange of the letters, it would seem to me that that might serve as additional evidence that there was a serious cost squeeze so far as the steel companies were concerned.

ANDY GLASS: New York Herald Tribune: When did the company's executive committee, sir, approve the price increase, and was the actual decision made before the signing of the labor agreement with Mr. McDonald?

MR. BLOUGH: The decision to have a price change?

MR. GLASS: Yes, sir.

MR. BLOUGH: It was authorized by the executive committee Tuesday of this week, Tuesday afternoon.

QUESTION: In view of your own statement about the urgent need to modernize in the face of foreign competition and the possibility that steel production will drop this summer, would you accept a public commission study of the merits of your cost-squeeze case, to report back by September 1st, we'll say.

MR. BLOUGH: Well, if I understand your question, you are asking whether we would accept some sort of an independent study as to the merits of our situation?

QUESTION: Yes, sir.

MR. BLOUGH: Well, I don't know under what circumstances that study would be made. There have been a number of considerations of that kind in the past. I really believe that it is quite important, in the kind of a society that we have, that individual companies in as highly competitive an industry as we have—and in every industry for that matter—should make their own price decisions.

ELMER MURRAY: Fairchild Publications: Mr. Blough, all the major industries, the textile industry, for example, has been the first to receive depreciation tax relief. And, of course, it has a great rate of obsolescence, and that may be explained. Could you tell us whether this decision would affect the textile industry adversely, and would you give us some estimates of what a heavy piece of capital equipment, like a loom, might cost?

MR. BLOUGH: Well, I am sorry, I am not very well versed on the textile industry. I know of nothing in connection with this steel matter that would, in any way, affect the textile industry.

QUESTION: Can you foresee any changes in the tax laws or any changes in depreciation allowances that would permit you to reconsider the price rise?

MR. BLOUGH: Well, all I can say is that this problem that is related to pricing is a continuous thing. All the factors that are involved in decision making are taken into account time after time over many years.

Now, if there are changes in the tax laws that are beneficial from the standpoint of depreciation which is certainly something that is needed, that would be, for sure, a factor that would be taken into account, of course.

DALLAS TOWNSEND: CBS News: The president seems to have been quite upset at the way he was informed. Do you feel, in the light of what has happened since then, that perhaps you could have given him a little more advance notice than you actually did?

MR. BLOUGH: Well, of course, I am quite concerned with the president's concern. In the kind of an economy that we have, I am not quite sure that it is feasible for anyone to be going to the White House to consider a price increase, even as important a one as this.

Now, I would like also to say that I am not sure it would have been the right thing to do under all the circumstances.

QUESTION: Mr. Blough, Joseph Senical, Radio Pulsebeat News: The President said, in his news conference, yesterday, the six-dollar per ton cost rise may add a billion dollars to our defense budget. Will you comment on this, please?

MR. BLOUGH: Yes, I'd be glad to! I am not sure how this estimate is made up. I am sure that, so far as direct steel sales to the Defense Department is concerned, that the total tonnage, estimating it any way you wish, to all subcontractors and everybody else, even if 1962 is an extraordinarily large year for purchases by the Defense Department, it could not amount to more than, say, three or three and a half million tons.

Now, those are the direct and indirect purchases, and that of course, would be, oh, something in the nature of twenty million dollars. And where the "billion dollars" comes from, I don't know, unless someone made a projection that this in some way would extend to other things.

QUESTION: Mr. Blough, sir . . .

MR. BLOUGH: I should point out, in connection with that, this price increase in steel isn't the only price increase that has occurred in the American economy. There have been quite a number, and there have been a number of price reductions, just the same as there have been price reductions in steel, so this question of movement of price is a very volatile thing.

QUESTION: Mr. Blough, Bill Steinbock, Westinghouse Broadcasting Company: Are we to understand, sir, from your statement, that the only possible way a price increase could have been averted would have been through a new contract that substantially reduced your employment costs and, if so, did you mention this to Mr. McDonald during the negotiations?

MR. BLOUGH: I think Mr. McDonald stated that he did not discuss prices with us, and that's certainly the fact! And we should not be discussing prices with the union.

QUESTION: Mr. Blough, in view of your stressing the need to modernize your plant and in view of the fact that United States Steel's capital expenditures dipped a hundred fifty million from 1960 to 1961, what plans—what plans does United States Steel have for the purchase of capital equipment and, particularly, of machine tools in the coming fiscal year?

MR. BLOUGH: I can answer that by saying, that will depend entirely on our earnings. If our earnings are up, we will purchase more capital goods.

QUESTION: And when will that decision be made, sir?

MR. BLOUGH: As the earnings improve!

QUESTION: Mr. Blough, would you care to comment on the statement made earlier today in a wire-service dispatch which quotes a government attorney saying this action by United States Steel has overtones of violations of two sections of the Sherman Anti-Trust Law?

MR. BLOUGH: Well, I don't know what he is referring to, if he made that statement, and I would question the statement.

QUESTION: Mr. Blough, Breidenham of Chicago Daily News: Sir, the justification for the price increase you have just given us, do you feel that applies equally to the increases announced by the other steel companies?

MR. BLOUGH: The other steel companies will have to speak for themselves.

QUESTION: Mr. Blough, in view of the comments made regarding foreign competition during the labor negotiations in which United States Steel made several major speeches regarding the element of foreign competition, could you explain it to those of us who don't understand these things very well just how you meet competition by raising your price?

MR. BLOUGH: You don't meet competition by having a facility produce a product that can't compete, cost-wise, with the imports. You've got to have a facility which can compete, cost-wise.

Your problem, as I have tried to explain, before, is to have enough to buy the machinery and equipment to make the kind of products, to make the quantities of them that will enable us, cost-wise, to get the business which, in turn, provides the employment.

QUESTION: Well, Mr. Blough—Murray Kempton, New York Post—some of us have a problem, here. United States Steel sells twenty-seven per cent of the total steel products; it seems unlikely that you would take the risk of raising your prices without some confidence that competing firms would do the same thing, and we are wondering, was there an understanding to this effect among the steel companies, or United States Steel's competition . . . ?

MR. BLOUGH: I can answer this, simply—no understanding of any kind!

QUESTION: Mr. Blough, Delaney, CBS News: Some significance has been attached to the fact that the steel companies, in 1960, did not raise prices under a Republican administration. Could you comment on that?

MR. BLOUGH: Well, I think you gentlemen can readily see that I do not know anything about politics! (*Laughter*)

QUESTION: Mr. Blough, Jim Bishop, Newsweek Magazine: Was the alternative of selective price increases over a long period of time ever considered in lieu of an across-the-board increase that you have gotten?

MR. BLOUGH: I am sorry, I didn't hear that question.

QUESTION: Were the alternatives of selective price increases over a long period of time ever considered in lieu of across-the-board increase?

MR. BLOUGH: The answer is "Yes."

QUESTION: Mr. Blough, Dave Jones, Wall Street Journal: You said that the decision to increase prices was made Tuesday, and yet you had indicated, quite a long time previous to that, that this cost situation that you faced was severe, and I am wondering what was different on Tuesday that wasn't different, that wasn't the situation, say, a week before that made you make this decision on Tuesday.

Was it the labor settlement? What bearing did it have?

MR. BLOUGH: There were—there are always a number of factors. Bear in mind that we had additional employment costs increases last October first—in fact, some in September! Now, we had experience with that quarter, final quarter of last year, and we had further experience this year. This was a decision that was—that involved many, many factors. There is no single factor that's involved in it.

QUESTION: Mr. Blough, is it correct—and, if it is, is it correct that foreign competition in steel in this country amounts to only about one week's production of the American steel industry and, if so, are you determined on this price issue to remain on a collision course until the government is forced to take some action that will be detrimental to all business?

MR. BLOUGH: Well, first, let me explain the second part of your question. This is an effort on our part to improve our cost/price relationship. At the moment, it's—I have no knowledge whether it will continue, or whether it won't continue because it will depend on what our competition does, primarily.

As to the foreign competition that you mention, the size of it, I made a little calculation some time ago. Bearing in mind that we are losing markets abroad, as well as having increased tonnage coming into this country, the total difference in, say, 1961 over our foreign competition picture in 1961 compared, say, with five or six years before would mean a difference of five or six million tons of ingots.

Now that, as has been previously stated, would represent as much as forty thousand jobs, so this question of competition from abroad is a very serious one.

QUESTION: Mr. Blough, you have said that you need increased profits in order to modernize your plants. Many corporations, however, including public utilities, when they want to expand their investment, issue new shares of stock, rather than tax the consumer through higher prices. Why do you choose this route, rather than seeking new equity capital in the corporation?

MR. BLOUGH: Well, let me say that we have been required to borrow, since 1958, including 1958, about eight hundred million dollars. Now, whether we borrow the money from investors and pay it back, or sell stock to them and try to improve our facilities from the proceeds of that stock is a question of management choice.

The point I make is that we have gone outside for a great deal of the money we have needed, and I must say that, if you borrow money, you've got to have the cost/price relationship that will give you enough profit upon which you pay the taxes and, after having paid the taxes, then you return the money that you borrowed—and we intend to do that.

one hundred fifty-seven

QUESTION: Mr. Blough, would the administration's pending proposal for accelerated depreciation on new-plant expenditure, on new plant, help you in your situation and, if so, to what extent?

MR. BLOUGH: Well, there are a number of features in the proposed tax bill, as you know, and if the bill should become law and if that particular feature is in the law then, to the degree that we'd be able to take advantage of it, it would be helpful.

But there are a number of "ifs" connected with any tax bill, including the amount that the new tax bill would cost us which, in turn, might offset to a degree or to a large degree any benefit that you would get from the particular provision that you spoke of.

QUESTION: Mr. Blough, you said in your statement that your letter of September thirteenth, anyone reading your letter of the thirteenth quote could not infer any commitment of any kind to do other than to act in the light of all competitive factors.

Could you tell us how the competitive situation for such steel products as structural steel or tin plate or concrete re-bars or wire and cable, which are suffering intense competition and which have had price weakness has changed since September thirteenth so as to require any price increase?

MR. BLOUGH: Well, since September thirteenth, I think the entire economy has improved. And, when the economy improves, the demand for steel increases, and one of the ways that the economy improves is when plant and equipment and all the other things that you talk about are purchased by people that have enough funds to purchase them. That's what helps build up the economy.

Now, in my view, the economy is in an improved condition today, compared to last September.

QUESTION: Mr. Blough, some representative from the government said your actions have been taken by quote wholly irresponsible and unjust handful of people—handful of steel executives. Would you reply to this?

MR. BLOUGH: I tried to explain, I thought, that our action was taken responsibly.

QUESTION: Mr. Blough—one thing, Mr. Blough: There is still talk, nevertheless, sir, in Washington, that the increase, coming as it did right on the heels of the labor pact, was timed to check expanded government influence in collective bargaining—in other words, that you acted politically, as well as economically.

Could you comment on that?

MR. BLOUGH: Well, I tried to explain, before, that I knew nothing about politics, and I have given you all the reasons why we had to change our . . .

QUESTION: There was no political motivation, sir?

MR. BLOUGH: I don't know what more I can say about that!

QUESTION: Thank you, sir!

MR. BLOUGH: I am sure that the answer to your question is, there is no possible, conceivable political motivation that I would know about. I don't believe that I would even know how to operate in that area.

QUESTION: Mr. Blough, sir, Yerkes of ABC News: Some expectation has been expressed that foreign competition and the substitution of other materials might force a revision in this price schedule. Has your Executive Committee made any estimate of the extent to which foreign competition and the substitution of other materials might cut into the gains that you anticipate from this price increase?

MR. BLOUGH: It was one of the considerations that we took into account.

QUESTION: And, have you determined . . .

MR. BLOUGH: And, you understand, there is nothing sure about being in business! The competitive situation, today, is a different one than it was yesterday, or tomorrow. The type of analysis that you are required to make all the time is a fundamental one. Under all the circumstances as we saw them, we thought this was the advisable thing to attempt to do.

QUESTION: Have you resolved what you think the net gain will be in a year's time?

MR. BLOUGH: I haven't any findings to give you.

QUESTION: Mr. Blough, as of the time of the start of this press conference, Armco and Inland Steel, the two major producers, as well as the larger, specialty producers, Ludnum and Crucible, have not yet raised prices. If they don't go along, how long can you stick to your price increase before you rescind? Or, would it affect you?

MR. BLOUGH: It would definitely affect us, and I don't know how long we could maintain our position.

QUESTION: Would it be a matter of days?

MR. BLOUGH: I wouldn't want to state a specific number of days.

QUESTION: But, if they don't follow within a reasonably-short period of time, you would expect to have to rescind, at least on some products, then?

MR. BLOUGH: It would make it very difficult for us!

QUESTION: Thank you!

QUESTION: Mr. Blough, Jack Allen, Mutual Broadcasting News: There have been a lot of people who fear this move by United States Steel may lead to further government regulation of all industry in the United States. Have you taken this into consideration when you made your price rise?

MR. BLOUGH: I saw no reason to think that any change in the price of steel would lead to government regulation. I still see no reason for that to happen.

QUESTION: There are those who may disagree, sir!

QUESTION: Mr. Blough, for the last—more than a year, the inflationary psychology which has gripped most of the economy has been pretty well dampened down. There is considerable fear, now being expressed, that your increase in the price of steel will touch off a new rise, a new wave of inflationary psychology.

How seriously did you take this national problem into consideration in making your decision?

MR. BLOUGH: Well, we considered everything connected with the price change, and I doubt very much whether what you say will happen, and I'd like to explain why.

Our problem in this country is not the problem with respect to prices; our problem is with respect to costs. If you can take care of the costs in this country, you will have no problem taking care of the prices. The prices will take care of themselves.

QUESTION: Mr. Blough, what—what is the labor cost of a ton of steel in 1958, and what is it in 1962?

MR. BLOUGH: Well, some tons of steel take certain number of man hours, and other tons of steel take other man hours and, if you are talking ingots, it's one thing and, in any event, I don't know that I have a specific figure to give you.

one hundred fifty-nine

But I am sure of one thing—that the employment costs which, at the end of 1961 were four dollars and ten cents an hour—now, that's the average, employment costs for the industry, about four dollars and ten cents an hour . . .

QUESTION: Per hour?

MR. BLOUGH: On a per-hour basis was . . .

QUESTION: What is the per ton?

MR. BLOUGH: It was up, I would say, forty cents over what it was in 1958.

QUESTION: On the hours.

QUESTION: (By Walter Cronkite, CBS News) You dispute the President's figures on unit costs and on profit. I wonder if you have any theory as to how he could have gotten hold of figures that, in your opinion, are so far off base.

MR. BLOUGH: No, I don't have any theory.

QUESTION: In that connection, if I could follow up, since there is a dispute between the President's economists and yourself, over the costs of a ton of steel, would you then agree to open up to a Congressional committee or some other public body, the data on which you arrive at your estimates, so that the public may make some judgment in this matter?

MR. BLOUGH: First of all, the discussion about the cost of a ton of steel means very little. If you look at the return which the industry has made year by year, or U. S. Steel has made year by year, you will see that, taking into account all the costs on all the tons of steel, the result has been going downhill.

Now, that ought to be a satisfactory answer.

QUESTION: It isn't sir, I'm sorry.

QUESTION: What reaction have you received from the metal working industry throughout the country to your price hike? Have you received any immediate reactions from the big two auto manufacturers, and the little three?

MR. BLOUGH: I personally have had no contact with the auto manufacturers, and I don't know what their reaction is as to others. I am sure there are many who would think that the price change was all right.

There would be some people who would feel otherwise. That happens every time.

QUESTION: Would you give us the details of this subpoena that was served on U.S. Steel, and whether or not U.S. Steel would welcome an investigation by the anti-trust subcommittee of the Congress and by the anti-trust division of the Justice Department?

MR. BLOUGH: Well, I don't have the details of the subpoena. In fact, I don't think I have seen it. But so far as welcoming investigations, I don't believe it is going to make much difference whether we welcome the investigations or not.

QUESTION: The first three months of the year are over. Could you give us any idea of what U.S. Steel's profits will look like for the first quarter, from here on, as compared to last year? And to what extent is that profit picture related to the action taken on prices?

MR. BLOUGH: Well, as I think you know, we only discuss our profits for the quarter at the quarterly meeting. But I can tell you that our profits for the first quarter are far from satisfactory.

QUESTION: Mr. Blough, you indicated that you didn't inform the President at an earlier date for certain circumstances. Will you please indicate what those circumstances are?

MR. BLOUGH: I am not sure that I follow your question, sir. Would you repeat it?

QUESTION: The question before was: Why wasn't the President acquainted with the situation of the price rise in an earlier period of time, and you said you didn't because of some circumstances. I wondered if you would indicate some of those circumstances.

MR. BLOUGH: I think you must have misunderstood me.

QUESTION: Then I will repeat the question: Why wasn't the President informed of this at an earlier date?

MR. BLOUGH: Why wasn't the President—

QUESTION: —informed of a price rise at an earlier date?

MR. BLOUGH: The President was informed of the price rise in what I hoped was as courteous a manner as could be devised under all the circumstances.

QUESTION: Were you surprised at the reaction of the President, Mr. Blough?

MR. BLOUGH: I think the answer to that should be that I was.

QUESTION: Can we pursue this a bit further: U.S. Steel Corporation, I think, prides itself on good public relations. Was any consideration, given prior to the announcement of the steel price increase, as to how to best cushion the effect on the public and on the government?

MR. BLOUGH: Mr. Worthington, as you recall, put a release out that stated the reasons for the price change, and I would like to commend to you all that you read that release. It contains a lot of valuable and interesting information, and I believe that some of the problems that we have had, with respect to the price change, would not have occurred if people had taken the time to read that release.

QUESTION: The President's Council of Economic Advisors' analysis shows that, while U.S. Steel profits have dropped the last four years, the profits of the other seven major steel companies have remained about the same. How do you explain that?

MR. BLOUGH: Well, I am not familiar with the analysis. I would like to say one other thing: We have been going about, almost an hour, and although I would be very happy to see you gentlemen on other occasions—and I hope that invitation doesn't get me into too many conferences with you—I would like at this point, if you are willing, to terminate the questions.

QUESTION: One other question, Mr. Blough.

MR. BLOUGH: I would like to finish, thank you.

(The press conference concluded at 4:30 p.m.)

one hundred sixty-one

EXHIBIT I *President Kennedy's Statement at News Conference* April 19, 1962

First, let me make it clear that this Administration harbors no ill will against any individual, any industry, corporation or segment of the American economy.

Our goals of economic growth and price stability are dependent upon the success of both corporations, business and labor and there can be no room on either side in this country at this time for any feelings of hostility or vindictiveness.

When a mistake has been retracted and the public interest preserved, nothing is to be gained from further public recriminations.

Secondly, while our chief concern last week was to prevent an inflationary spiral, we were not then and are not now unmindful of the steel industry's need for profit, modernization and investment capital.

I believe, in fact, that this Administration and the leaders of steel and other American industries are in basic agreement on far more objectives than we are in disagreement.

We agree on the necessity of increased investment in modern plant and equipment.

We agree on the necessity of improving our industry's ability to compete with the products of other nations.

We agree on the necessity of achieving an economic recovery and growth that will make the fullest possible use of idle capacity.

We agree on the necessity of preventing an inflationary spiral that will lead to harmful restrictions on credit and consumption.

And we agree on the necessity of preserving the nation's confidence in free private collective bargaining and price decision, holding the role of government to the minimum level needed to protect the public interest.

In the pursuit of these objectives we have fostered a responsible wage policy aimed at holding increases within the confines of productivity gains. We have encouraged monetary policies aimed at making borrowed capital available at reasonable cost; preparing a new transportation policy aimed at providing increased freedom of competition at lower cost; proposed a new trade-expansion bill to gain for our industry increased access to foreign markets; proposed an 8 per cent income tax credit to reward investment in new equipment and machinery, and proceeded to modernize administratively the Treasury Department's guidelines on the depreciable lives of capital assets, and finally taken a host of other legislative and administrative actions to foster the kind of economic recovery which would improve both profits and incentives to invest.

I believe that the anticipated profits this year for an industry in general—and steel in particular—indicate that these policies are meeting with some measure of success.

And it is a fact that the last quarter of last year and I think the first quarter of this year will be the highest profits in the history of this country; and the highest number of people working and the highest productivity, so that while there are serious economic problems facing us, nevertheless, I believe that progress is being made and can be made and must be made in the future.